Small Far
Big Change
Scaling up impact in smallholder agriculture

Edited by David Wilson, Kirsty Wilson and Claire Harvey

Order Form

How to order

To receive a 10% discount, order from www.developmentbookshop.com

☏ +44 (0) 1926 634501

✉ complete the order form below and return to:
Practical Action Publishing, Schumacher Centre for Technology and Development, Bourton on Dunsmore, Rugby, Warwickshire, CV23 9QZ, UK.

Postage & Packaging:
UK – allow 4-5 days.
Europe – allow 6-15 days.
Rest of the World Allow – 28 days, or up to 180 days to remote areas.

Payment Details (please tick payment method)

I enclose payment of total cost [   ]

(cheques should be made payable to Practical Action Publishing)

[ ] **Please send me a proforma invoice**

[ ] **Please charge my credit card** Mastercard/Visa/Maestro

Card No. [   ]

Valid from [   ]   Expiry date [   ]

Issue number (Maestro only) [   ]

Security number (last 3 figures on the security strip) [   ]

Signature [   ]

Title and ISBN	Price	Quantity	Amount
Small Farmers, Big Change ISBN 978 1 85339 712 7	£14.95		
Cost of Postage (see below)			
Grand Total			

Cost of Postage
UK, add 15%. min. £2, **Europe**, add 20%. min. £2,
Rest of the world orders up to £100, standard delivery, add 25%. min. £2, **Rest of the world orders over £100**, standard delivery, add 20%. min. £2,
Rest of the world priority service add 40%.
Please contact us for a quote on courier of airfreight services

Cardholder's Details

Name	
Address	
Postcode	Country
Telephone	Email

Delivery Address (if different from cardholder's details above)

Name	
Address	
Postcode	Country
Telephone	Email

Any profit supports the work of Practical Action, Reg. Charity No. 247257.   *Thank you for your Order!*

For Practical Action Publishing's full catalogue or any other queries please email publishinginfo@practicalaction.org.uk

Publication dates, prices and other details are subject to change without prior notice. 1998 Data Protection Act.
Practical Action Publishing will NOT pass your details on to a third party.

1 2 3 4 5 6 7 8 9 10

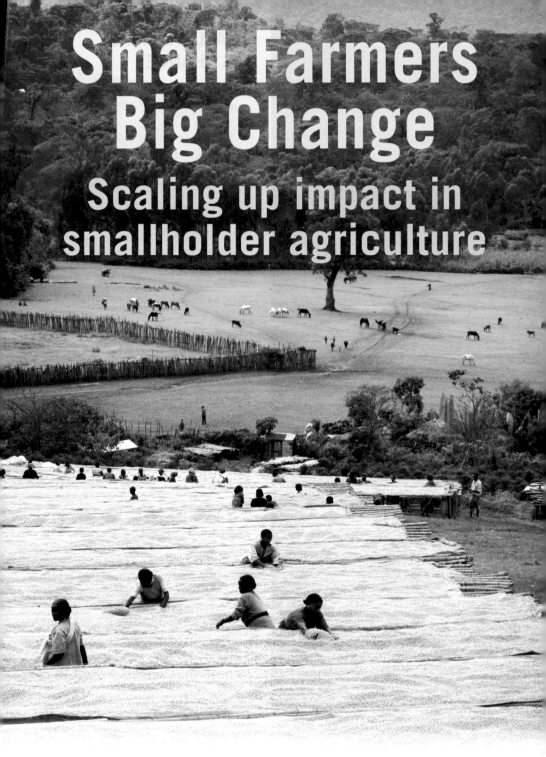

Small Farmers Big Change

Scaling up impact in smallholder agriculture

Edited by David Wilson, Kirsty Wilson and Claire Harvey

Order Form

How to order

To receive a 10% discount, order from www.developmentbookshop.com

☏ +44 (0) 1926 634501

✉ complete the order form below and return to:
Practical Action Publishing, Schumacher Centre for Technology and Development, Bourton on Dunsmore, Rugby, Warwickshire, CV23 9QZ, UK.

Postage & Packaging:
UK – allow 4-5 days.
Europe – allow 6-15 days.
Rest of the World Allow – 28 days, or up to 180 days to remote areas.

Payment Details (please tick payment method)

I enclose payment of total cost [   ]

(cheques should be made payable to Practical Action Publishing)

☐ **Please send me a proforma invoice**

☐ **Please charge my credit card** Mastercard/Visa/Maestro

Card No. [   ]

Valid from [   ]   Expiry date [   ]

Issue number (Maestro only) [   ]

Security number (last 3 figures on the security strip) [   ]

Signature [   ]

Title and ISBN	Price	Quantity	Amount
Small Farmers, Big Change ISBN 978 1 85339 712 7	£14.95		
Cost of Postage (see below)			
Grand Total			

Cost of Postage
UK, add 15%. min. £2, **Europe**, add 20%. min. £2,
Rest of the world orders up to £100, standard delivery, add 25%. min. £2, **Rest of the world orders over £100**, standard delivery, add 20%. min. £2,
Rest of the world priority service add 40%.
Please contact us for a quote on courier of airfreight services

Cardholder's Details

Name	
Address	
Postcode	Country
Telephone	Email

Delivery Address (if different from cardholder's details above)

Name	
Address	
Postcode	Country
Telephone	Email

Any profit supports the work of Practical Action, Reg. Charity No. 247257.   Thank you for your Order!

For Practical Action Publishing's full catalogue or any other queries please email publishinginfo@practicalaction.org.uk

Publication dates, prices and other details are subject to change without prior notice. 1998 Data Protection Act.
Practical Action Publishing will NOT pass your details on to a third party.

1 2 3 4 5 6 7 8 9 10

Small Farmers,
Big Change

Praise for this book

'It is not a lack of large-scale farms that lies behind the disappointments of agricultural development in the developing world, but a lack of conditions to allow small farmers to fulfil their potential. The history of modern farming reveals numerous episodes where there have been remarkable spurts of technological innovation and growth in agricultural output coming from small farms. The cases in this important book show that when small farmers are given the right incentives to produce more and the means to do so, they invest, innovate and respond to opportunity.'

Dr John Thompson, Research Fellow, Institute of Development Studies, UK and Joint Director, Future Agricultures Consortium

Small Farmers, Big Change

Scaling up impact in smallholder agriculture

Edited by David Wilson, Kirsty Wilson
and Claire Harvey

Published by Practical Action Publishing Ltd in association with Oxfam GB

Practical Action Publishing Ltd
Schumacher Centre for Technology and Development
Bourton on Dunsmore, Rugby,
Warwickshire CV23 9QZ, UK
www.practicalactionpublishing.org

Oxfam GB,
Oxfam House, John Smith Drive,
Oxford OX4 2JY, UK

ISBN 978 1 85339 712 7

© Oxfam GB, 2011

A catalogue record for this book is available from the British Library.

The contributors have asserted their rights under the Copyright Designs and Patents Act 1988 to be identified as authors of their respective contributions.

Since 1974, Practical Action Publishing (formerly Intermediate Technology Publications and ITDG Publishing) has published and disseminated books and information in support of international development work throughout the world. Practical Action Publishing Ltd (Company Reg. No. 1159018) is the wholly owned publishing company of Practical Action Ltd. Practical Action Publishing trades only in support of its parent charity objectives and any profits are covenanted back to Practical Action (Charity Reg. No. 247257, Group VAT Registration No. 880 9924 76).

Oxfam is a registered charity in England and Wales (no 202918) and Scotland (SCO 039042). Oxfam GB is a member of Oxfam International.

Cover photo: Arabica coffee beans are turned and graded on the drying installations at a coffee processing facility supported by the Oromia Coffee Farmers Cooperative Union (OCFCU), Ethiopia. OCFCU, established in 1999, helps farmers to implement fair trade and organic working practices in the renowned Yirgacheffe coffee-growing region. Credit: Sven Torfinn, Panos Pictures

Indexed by Liz Fawcett, Harrogate, North Yorkshire
Typeset in Stone Serif by Bookcraft Ltd, Stroud, Gloucestershire
Printed by Hobbs the Printers Ltd
Printed on FSC 100% post-consumer waste recycled paper.

Contents

Figure

Tables

Boxes

Acknowledgements

In 2005, Oxfam launched the Global Agriculture Scale Up Initiative (GASUI) with the aim of empowering smallholder farmers to improve their own livelihoods. In 2009, after a pilot phase of implementation, Oxfam staff documented programme learning on achieving impact at scale from the four core countries in GASUI (Ethiopia, India, Honduras and Tanzania) to share with wider Oxfam programmes supporting small-scale agriculture.

The 'Small Farmers, Big Change!' workshop, held in Oxford on 12–14 May 2009, brought together 23 participants from 17 country and regional programmes and three Oxfam affiliates (Oxfam GB, Oxfam America, Oxfam India) together with external speakers and global advisory staff to discuss this evidence alongside learning from other countries. This book is an outcome of that learning process.

Sally Baden, currently Senior Adviser, Agriculture and Women's Livelihoods, Oxfam GB, designed and facilitated the 'Small Farmers, Big Change!' event, provided overall editorial guidance on the selection of the chapters for this book, and feedback on earlier drafts, in addition to co-authoring the overview chapter. Kirsty Wilson coordinated the event, with support from Olugbenga Akinyooye and Francoise Kambabazi.

Among Oxfam's global advisory staff who contributed both to the event and to commenting on drafts of chapters in this book were: Lea Borkenhagen, David Bright, Constantino Casasbuenas, Tim Chambers, Alan Doran, Eric Hazard, Thalia Kidder, Sally King, Tim Mahoney, Hugo Sintes, Shahid Zia and Amit Vatsayan. There is a long list of other Oxfam staff and external experts who commented on chapters and inputted into the development of the book. We gratefully acknowledge these contributions.

Oxfam gratefully acknowledges the financial assistance of the Accenture Foundation in supporting the learning event and related research and documentation.

Contributions to the 'Small Farmers, Big Change!' event are also acknowledged from Mike Albu, Practical Action, Catherine Martin, DFID, Dr Jim Sumberg, IDS, Dr John Thompson, IDS, and Dr Bill Vorley, IIED.

Emma Donne, research assistant on *Small Farmers, Big Change,* is currently conducting research with Fairbridge and TMP Worldwide.

Katie Allan and Abigail Humphries Robertson were project managers for *Small Farmers, Big Change.* Katie is now Information and Communication Officer, Greater Horn of Africa Rainwater Partnership (GHARP)/Kenya Rainwater Association (KRA). Abigail is Editorial Project Manager, Oxfam GB.

About the authors

David Wilson, Kirsty Wilson and Claire Harvey co-edited *Small Farmers, Big Change*. David is a freelance editor/consultant/journalist working mostly in the development/humanitarian field and international trade and business. Kirsty was Programme Resource Officer – Sustainable Livelihoods, Oxfam GB at the time of writing and is now still working for Oxfam as the Project Manager for the Africa Climate Change Resilience Alliance (ACCRA) in Ethiopia. Claire is Communications Manager for content development in the Policy and Practice Communications team, Oxfam GB.

Small farmers, big change: Lessons from Oxfam's agricultural programmes
An overview

Sally Baden and Claire Harvey

Introduction

Improving smallholder agricultural livelihoods is essential for achieving large-scale poverty reduction and growth. Almost 80 per cent of the world's 925 million hungry people live in rural areas, and most depend on agriculture as their main source of income and employment. Approximately half of these are smallholder farmers.[1] Overall, smallholder farmers constitute 1.5 billion of the three billion people living in rural areas and 87 per cent of all farmers in developing countries.[2] Of the one billion poor people living in rural areas, most rely mainly on agriculture for their incomes.[3] Equally, agriculture is key to achieving broad-based growth, especially in low-income developing countries.[4] Investing in smallholder agriculture ensures that this growth is inclusive, pro-poor, and environmentally sustainable. Under certain conditions, it can also be more efficient than large-scale agriculture.[5]

In many developing countries, and particularly in sub-Saharan Africa, women provide a high proportion of agricultural labour, particularly in food production, processing, and marketing.[6] Women work as subsistence farmers, small-scale entrepreneurs, unpaid workers on family farms, or casual wage labourers, and often they play several of these roles. Because they are also responsible for the majority of caring and household tasks, women's working hours are often longer than men's, limiting their scope to engage in new opportunities. At the same time, women farmers also have more restricted access to land, credit, and training than their male counterparts.[7] Where women are involved in marketing the crops that they grow, it is often in small volumes and in the least profitable markets, and often does not lead to significant increases in income. These factors not only limit women's benefits from smallholder agriculture but, crucially, also reduce overall agricultural productivity by as much as 10 per cent.[8]

Oxfam's Global Agricultural Scale Up Initiative (GASUI) was launched in 2005 with the aims of reducing poverty for millions of smallholder farmers, particularly women; driving economic growth by linking farmers into new and wider market opportunities; and advocating for increased donor, government, and private sector investment to support smallholder farming. For Oxfam, 'scale-up' refers to increasing both the reach and quality of Oxfam's agricultural programming and thereby demonstrating to governments, donors, and development actors how agriculture can contribute to economic development and poverty reduction. The initiative's main strategy is to empower smallholder farmers to organize and engage effectively and equitably in agricultural markets and value chains.

This overview precedes eight chapters based on Oxfam's programme experience. The chapters draw on learning from GASUI and on wider learning from Oxfam GB's agricultural markets-based programming, as well as from programmes implemented by Oxfam India.

They were originally written for a global learning event in May 2009,[9] and have been further developed to show different pathways for achieving scale in a variety of contexts. The case studies explored in the chapters demonstrate how linking small producers to markets and value chains and enabling them to engage in policy processes can help to improve livelihoods and can be a catalyst to wider, longer-term change. In developing such linkages, Oxfam programmes are working with private companies as well as with farmer organizations to develop new business models, which maximize the benefits of market engagement for smallholder farmers, especially women, and contribute to economic growth. Programmes are also developing innovative methods of service delivery to small farmers, in collaboration with financial institutions and other service providers. Such innovations have the potential to be replicated by others, beyond Oxfam's programme interventions.

These case studies are intended as a learning resource for development practitioners and professionals working in the field of smallholder agriculture, particularly those promoting rural women's rights and women's economic leadership. The different case studies presented offer an understanding of the varied livelihood options and risks faced by diverse groups of smallholders and possible strategies for managing these, in the context of globalized markets and a changing policy environment. The analysis and recommendations will be helpful to those involved in developing and delivering agricultural livelihoods programmes, especially those thinking about how to 'scale up' interventions. They will also be of interest to people working in business in the agricultural sector with an interest in linking to small farmers in developing countries, as well as to researchers interested in agricultural development. They provide evidence and examples useful for governments, donors, civil society, and private sector organizations on the potential of investment in smallholder agriculture for contributing to poverty reduction, enterprise development, and economic growth.

The changing context of smallholder agriculture

From the 1980s up to 2006, government and donor support to agricultural investment declined overall, and support to smallholder agriculture in particular, in most developing countries.[10] Policies of liberalization in the agricultural sector during this period reduced government support to agricultural production and marketing and to rural infrastructure and services. The private sector has not effectively filled the gap left by the withdrawal of the state. At the same time, the power of large agribusinesses in the sector has grown, with the globalization of agricultural value chains and the deregulation of markets.

The food price crisis, which dramatically hit global markets in 2008, underscored the legacy of this underinvestment and brought agriculture back to the forefront of the development debate. Concerns about the security of food supplies in the face of growing urban populations and of climate change have led to a renewed focus on efforts to improve agricultural productivity and growth, to new commitments to agricultural investment, and to growing interest in more sustainable, low-carbon production systems. There is now an emerging consensus that, without significant increases in investment in agriculture, and in small-scale farming in particular, the Millennium Development Goals for poverty and hunger reduction cannot be reached.[11]

Developing rural food production can not only help to address the rural/urban income gap, but can also provide food for growing urban populations in the face of potential future food crises. The question is no longer about whether to invest in smallholder agriculture, but in what, where, and how. While the case for increased investment in agriculture is widely accepted, and in some cases has translated into significant commitments of funds globally and nationally, translating this into investments that benefit the majority of small farmers, particularly women, remains a huge challenge.

These chapters are about approaches to promoting the knowledge, technologies, policies, and institutional frameworks that enable agricultural markets to function in ways that benefit both poor rural people and the wider economy. Significant change is required to create strategies that are not only technically 'accurate' and able to address complex problems, but also effective in achieving widespread impact and improved life chances for both women and men smallholder farmers. Equally, in response to the growing risks to viable livelihoods associated with climate change, the interventions described in the chapters aim to equip smallholder farmers for the future by identifying sustainable opportunities, as well as viable strategies, for their engagement in markets.

Oxfam's work in smallholder agriculture

Oxfam's work in supporting smallholder agriculture has for many years been a major focus of efforts to promote poor people's rights to a sustainable livelihood. Work on sustainable livelihoods is part of a broader, integrated rights-based approach to addressing poverty and suffering. This includes promoting farmer organization and advocacy at all levels, from local to global, to influence the policies and institutions that shape smallholders' livelihood opportunities, with an emphasis on supporting the political participation and leadership of women. Equally, rural areas have long suffered from lack of investment in infrastructure and essential basic services. Such lack of investment has not only denied poor rural people and their families their basic rights, but has limited the capacity of poorer farmers, and particularly women farmers, to play a productive role and to participate fully in development. Poor rural people are also often the most vulnerable to crises, suffer disproportionately from the impacts of climate-related risks, and are increasingly affected by conflicts or outside interventions that dispossess them of their sources of livelihood. Promoting their rights to essential services and reducing their vulnerability to shocks is also a core part of Oxfam's work.

Since 2003, recognizing the shift of power in the agricultural sector towards large-scale business, Oxfam's work with small farmers has focused on strategies to improve their level of organization and their negotiating position in markets, in order to increase their share of agricultural incomes. This has complemented longstanding work on strengthening smallholders' access to and control over assets, recognizing the increasing importance of cash incomes to poor people's livelihoods. The GASUI initiative has reflected and built on this change, with its emphasis on empowering smallholders to improve their own livelihoods by gaining greater 'power in markets' that increases their ability to access and effectively participate in markets and engaging in policy processes and partnerships with private sector organizations. In addition, in view of the decline in investment in agriculture in general and in smallholder agriculture in particular over many decades, GASUI has also aimed to build a body of evidence and experience to demonstrate the importance of public, as well as private, investment in small-scale agriculture as a means to deliver longer-term poverty reduction and economic growth in different contexts.

Since 2005, GASUI has been implemented in three core countries – Honduras, India, and Ethiopia – with a fourth country, Tanzania, joining in 2007. Partly informed by learning from GASUI, a growing number of other Oxfam country and global programmes – some of which are reflected here – have also been working to support efforts to influence policy, engage in markets and value chains, and promote relevant services to strengthen small producers' livelihoods. As well as having a greater focus on linking small producers to markets, GASUI has represented a shift in Oxfam's approach to livelihoods programming, away from asset provision or service delivery in projects to one of facilitating wider processes of change in order to achieve a wider impact. Table 0.1 below characterizes the shifts in Oxfam's thinking about how to achieve scale in agricultural programmes that GASUI has been instrumental in promoting.

'Scaling up' impact in smallholder agriculture does not mean simply multiplying the number of projects or investing in bigger projects. Rather, it is about doing things in different ways, for example by forming alliances with other actors to leverage greater overall investment. In fact, reliance

Table 0.1 Oxfam's new focus in livelihoods programming

Original thinking on 'scale'	New thinking on 'scale'
Scale achieved through increasing programme numbers: countries, companies, projects, beneficiaries	**Scale achieved through increasing influence** through innovation, strategic partnerships, alliances, knowledge sharing
Scale increases proportionally to size of programme	**Scale achieved exponentially through self-spreading** of ideas, practices, services beyond Oxfam's direct reach
Scale is reached by increasing Oxfam's programme funding	**Scale is reached by leveraging wider investment resources** by and to smallholders and supporting service providers
Ad hoc, opportunistic market and company engagement	**Systematization of analysis, tools, and processes** for identification of markets, products, companies based on potential for scale, value added, inclusion, and specifically women's economic leadership

primarily on donor or INGO resources is likely to lead to 'unsustainable' interventions.

Ultimately, achieving scale involves employing innovative and self-sustaining strategies to achieve profound change with the potential for positive impacts on the lives of large numbers of poor people. Interventions can be self-sustaining if they are designed in collaboration with other key actors such as producer organizations, processing companies, district authorities, and providers of financial services. These actors, rather than Oxfam or other NGOs, are the drivers of processes of change, and this requires Oxfam staff and programmes to 'let go' of the process.

Diverse 'scaling' mechanisms exist within the agricultural sector. These include the farmer-to-farmer spread of new technologies or production practices; franchising, or adoption of new business models by companies; institutional and policy changes across municipalities, regions, or countries; and increased private or public sector investment in innovative and effective models of service delivery to poor rural people, such as mobile banking and the demand-driven spread of new services and technologies.

The role of Oxfam and other NGOs in these processes is to act as innovators and facilitators, working with multiple stakeholders. Such ways of working encourage the emergence of new types of farmer organizations and networks, strengthen the links between small farmers and private companies, and enable the formation of new alliances to influence policy and investment in ways that support small producers. NGOs can also play an important role in this context in promoting communications and learning by different stakeholders.

Pathways to leveraging change in smallholder agriculture

Through GASUI, Oxfam has promoted a 'market development' approach to agricultural programming. In this approach, key 'pathways' to leveraging wider change in smallholder agriculture (see Figure 0.1 below) are:

- Support to producer organizations and enterprise development;
- The facilitation of links with other market actors;
- Enabling wider access to key service providers, such as finance, transportation, and training;
- Improvements in the 'enabling environment' for smallholder agriculture, including e.g. trade policies, policies on women's land rights, and investment in infrastructure.

Figure 0.1 Pathways to scale

These case studies demonstrate examples of the successful application of a number of these strategies, which are now being 'scaled up' or replicated.

Supporting farmer organization underpins the other strategies and is an area in which Oxfam can demonstrate many successes, as shown here in examples from India, Colombia, and Mali. As described in chapter two, 'Strength in numbers', Indian fishing communities in Madhya Pradesh established village co-operatives and formed a federation that has given them a strong voice. This helped their campaign to persuade the state government to revise its fisheries policy and introduce a new law that protects the rights of traditional fishing communities. Chapter three, 'Leading by example', shows how networks of small farmers, NGOs, and other organizations in Colombia worked together to build a successful evidence-based advocacy strategy to influence municipal authorities to allow small agricultural producers access to urban food markets. Influencing of policy and investment decisions also took place in Honduras. Chapter one, 'The voice of many', describes how an alliance of rural civil society groups, including small farmers, worked with municipal officials to develop joint plans for the use of funds to implement a regional Poverty Reduction Strategy (PRS), in ways that responded to rural people's needs.

Farmer organization is a growing area of interest among development actors. Increased budgets for agriculture mean that successful, yet often young and fragile, farmer organizations are being sought out by donors requiring partners with the capacity for grassroots delivery and by companies seeking new suppliers. Although this is positive, it brings with it the danger of producer organizations being overloaded with responsibilities when their technical capacities and governance structures are weak. Oxfam's learning highlights the need to be clear about the different types of organization and support required for advocacy, enterprise development, and service delivery and to ensure that support to farmer organizations is cost-effective, sustainable, and replicable. It also highlights the fact that specific targeted support, such as literacy and technical training, and flexibility in ways of organizing – including separate women's activities and organizations – are often required to ensure that women as well as men can participate in and benefit from the opportunities offered by organization of farmers (see chapters two, four, five, and eight for examples).

Facilitating smallholder engagement in markets and value chains has taken different forms depending on the context and type of market. Chapter four, 'Engaging smallholders in value chains', shows how in Ethiopia linking farmers to the private company Ambrosia Ltd has enabled beekeepers to access training services – and international markets – through a joint Oxfam/ private sector investment. For farmers in Sri Lanka, a relationship with agri-based company Plenty Foods has offered market opportunities and enabled them to increase their incomes, while also benefiting the company's growth rates, as shown in chapter six, 'Growing partnerships'. Chapter five, 'Power to producers', describes how local farmers' associations in Haiti came together to supply a network of dairies with milk. Through this producer-led enterprise model, they were able to access technical and material support and benefits, and increase their incomes.

Innovation in services in Sri Lanka, where links were facilitated between financial service providers and rural enterprises, has enabled the enterprises to grow significantly and has provided models for replication, as described in chapter seven, 'Bridging the gap'. Chapter eight, 'Effective co-operation', shows how building the capacity of cotton producers' co-operatives in Mali has helped them to provide services to their own members. It has also helped them to build partnerships with lending institutions that are viable in the long term and has increased the participation of women farmers in the running of cotton co-ops.

Balancing scale with diversity and risk

The market development approach has tended to focus on those small producers who are capable, with support, of organizing and making use of market opportunities, rather than the most vulnerable or food-insecure farmers. Investments are made in existing rural activities to increase farmers' returns rather than activities that maintain livelihoods at subsistence level.

Nevertheless, it is important that market-based interventions are able to be inclusive of different social groups. Oxfam's development and use of tools such as 'gendered market selection' and 'gendered market mapping' (see Box 0.1) have resulted in interventions that are designed specifically to promote rural women's economic leadership, such as that described in chapter four on Ethiopia and chapter five on Haiti.

Even for small farmers who have some assets, livelihoods are highly vulnerable to risks from weather-related events such as drought. Some small producer programmes have developed links between market-based activities and social protection programmes to prevent the depletion of assets or to reduce vulnerability. For example, in India fishponds have been affected by successive droughts. In the programme described in chapter two, the participation of fisherfolk in labour programmes under the government's National Rural Employment Guarantee Act has enabled better pond maintenance. At the same time, advocacy on policy has ensured that fisherfolk are compensated for drought-related losses and are also able to use drought-affected pond beds to diversify into other economic activities.

Lessons from scaling up impact in smallholder agriculture

Smallholder agriculture can be the most effective route to bring about both economic growth and poverty reduction, and investment in this sector offers opportunities for increasing the resilience of small farmers to disasters. However, different and flexible strategies to achieving scale in smallholder agricultural programmes are required depending on the political, economic, and social context and on the nature of the specific markets and actors involved.

One lesson that emerges is that the potential and viability of smallholder agriculture as a strategy for sustainable livelihoods and development are not uniform across countries or types of market; neither are they equally relevant for all social groups. In countries where large-scale agriculture is dominant, alternatives might instead be to provide support to agricultural wage labourers, other forms of rural employment, or migration, or to work with large-scale farmers and agricultural companies to improve their employment practices.

Also, both the opportunities and the risks for small farmers vary with the type of market. When developing programmes, some of Oxfam's partners in country apply systematic criteria for selecting markets, while others are more opportunistic. Currently Oxfam's agricultural programmes work across a wide range of agricultural markets, including local food markets (e.g.

Box 0.1 Gendered market mapping[12]

Gendered market mapping is the process used to identify economically viable products and market opportunities that are profitable for women. The market map is a comprehensive visual representation of the different steps and actors in a particular value chain. It also shows the external factors affecting the chain (e.g. infrastructure, natural or policy environment) and the market services (e.g. transport, finance, information, extension services) needed to enable the chain to function.

A market map becomes gendered when:

- We include sex-disaggregated data – i.e. how many women and men participate in certain levels and positions in market chain(s) and how many women and men are benefiting from each provider of market services.
- We pay special attention to the infrastructure and services crucial for women because of their household and social roles (water, health, energy, safe transport, extension services run by or for women).
- We identify throughout the map the policies, practices, ideas, and beliefs that (dis)enable women's economic leadership – of traders, co-operatives, buyers, policy-makers, service providers, etc.

A gendered market map is useful for designing interventions because it helps us identify how women can move up the market chain. We can identify opportunities to upgrade women's position in the market system, in both the short and long terms. Market mapping is a tool that encourages us to talk to all buyers, producers, and service providers in a certain sector, and understand how the market looks to them. The visual map enables all actors to see themselves in relation to others in the sector, and helps show why actors might collaborate or change. It can raise their awareness of how gender inequalities may be affecting the quality or quantity of production and the timely delivery of produce to buyers. Women, especially, see themselves more clearly. It stimulates them to think beyond their existing activities and to see where new or more profitable market opportunities might exist.

horticulture in Honduras, fish in India, dairy in Haiti); high-value markets for semi-processed products in collaboration with processing companies; and 'niche' export markets (e.g. organic coffee and honey in Ethiopia and organic cotton and shea butter in Mali).

Local markets are easy to enter but may have limited potential for improving livelihoods at scale, so there is also a need to link farmers with growing urban, regional, and 'higher-value' markets. Entering higher-value markets may require improvements in quality or other standards and so may tend to exclude those with fewer assets and capacities, unless specific training or support is provided. Over-dependence on external markets is also risky, especially given the increasing volatility of commodity prices and demand. In particular, 'locking in' producers to one buyer is not desirable. Ideally, market development programmes should provide opportunities across different markets (local/national as well as international) to maximize scale and inclusion and to minimize risk.

Agricultural scale-up can offer opportunities for the economic empowerment of rural women. However, the capacity of farmers to adapt and change or to access, understand, or adopt new ideas is affected by their self-confidence, education, and social position. Gender, as well as ethnic or caste background, can be very significant here. Poor women face particular challenges to their participation in markets. These include lack of assets and asset security; lack of time to engage in productive work because of their unequal share of household responsibilities; persistent inequalities in access to resources and decision-making in the household; and legal discrimination and other barriers.

These challenges can be addressed by appropriate public investment – for example, through asset redistribution or infrastructure development to enable more sustainable, equitable outcomes for smallholders. They can also be addressed by proactive strategies to promote women's economic leadership in agricultural markets.

In Ethiopia (chapter four), beekeeping is one of the most sustainable livelihood options for landless people, and women generally do not own land. Growing numbers of female beekeepers have learned how to manage improved hives and beekeeping tools and equipment. In Haiti (chapter five), women's associations have given their members training on gender equality, which has helped them as producers and in the home. Many women now know that men need not be the sole decision-makers, and that they themselves should also play an important part in this process. In all cases, proactive strategies are required to ensure the participation of women in market-based programmes, where there are social and cultural barriers to this.

From the experiences described in the chapters, Oxfam has learned that movements for policy change are more likely to succeed where there is strong community organization and ownership, as demonstrated in the work undertaken by fishing communities in India (chapter two). Women's engagement in producer organizations, as well as in advocacy and campaigning activities, increases their confidence more broadly as market actors. The programme in Honduras (chapter one) has established over 90 savings and loan community funds with over 2,000 beneficiaries, 48 per cent of whom are women. Women's groups were also involved in the advocacy campaign that leveraged change in policy and investment at the national level. In Colombia (chapter three), one of the programme's main objectives was to support small-scale rural producers, especially women, by strengthening their role in the rural–urban food supply chain. Women played an active role in campaigning activities and in building public support for a food supply plan in Bogotá. Most women who had been engaged in advocacy activities felt more confident about marketing and selling their produce as a result.

In engagements with the private sector, as well as benefits for farmers there needs to be a strong business case for companies to engage with smallholders. This was the case with Plenty Foods – described in chapter six – where a 'win–win' arrangement for both company and producer groups has enabled small farmers, especially women, to increase their market opportunities and their incomes, while the company has secured its supply base and has achieved an annual growth rate of 30 per cent through shifting from working with individual farmers to organized groups of small farmers.

Future perspectives

The context in which we are working in 2011 is significantly different than when GASUI began in 2005, and Oxfam's work must adapt to this changing context. Demand for food is rising with growing populations and urbanization, and there is increasing conflict over limited resources. There are now greater shortages of food and water and increased climate instability. Climate change is undermining the long-term viability of agricultural livelihoods in some areas. Equally, with the increasing globalization of food and agricultural markets, global corporations have more power in these markets. There is increased volatility in commodity markets, with the risk of further marginalization and increased vulnerability of smallholders.

However, the renewed interest in agriculture from governments, donors, and private companies and recognition of the importance of small-scale agriculture for pro-poor economic growth provide a huge opportunity. Oxfam's next challenge is to maximize this opportunity by continuing to demonstrate how to most effectively leverage wider change for smallholder agriculture. We will continue to improve our understanding of small-scale agriculture in livelihoods by revisiting strategies for increasing our impact in different contexts. In doing so, it is essential that we identify and clearly state how we expect change to happen within our programmes, what our role is in stimulating that change, and how we define 'scale' as a matter of depth, not just breadth.

Building on our work so far, Oxfam is continuing to innovate in collaboration with both global and domestic private sector companies, as well as with producer-led enterprises, to develop and scale up new business models that can integrate small producers into supply chains in a sustainable and equitable way.[13] Women's economic leadership in agricultural markets is critical to ensuring equitable outcomes from smallholder development, as well as to ensuring food security at household and community levels.[14]

Development actors and private companies and service providers need effective 'models' for engaging directly with women producers – whether in single-sex or mixed organizations – that are economically sustainable as well as equitable. Actors on the ground also want to know when to promote women organizing separately and when to work via mixed groups, and what different forms of organization and strategy are most effective in different kinds of market and in enabling access to different services. There also remains a longer-term challenge to ensure that, as commercial opportunities develop in key markets, women are able to maintain and build their position within market systems, as well as to consolidate their bargaining power within households and communities.

We also want to enable small producers to be able to work effectively in a future context that may be very different from that of today. This means supporting them to manage resources and risks differently and to adapt to new 'shocks', bearing in mind that both climate change and globalization are changing future risk profiles. Oxfam needs to better understand traditional systems of risk management and to strengthen these, as well as to develop innovative mechanisms for sharing risk among actors along the value chain. It also needs to ensure that small producers, particularly women, have access to information about the likely impacts of changes in weather patterns and that they are supported to adapt production practices and techniques, for example by using new seed varieties. Equally, Oxfam needs to work with smallholders to assess both current market-based livelihoods and new market opportunities against likely future climate change impacts and to identify diverse livelihood strategies that will reduce vulnerability to these impacts.

Finally, alongside this, Oxfam will develop advocacy and campaigning activities to ensure that growing competition for scarce resources in the coming years does not further disadvantage small producers and that governments and companies, as well as NGOs, invest in positive, productive, and sustainable ways in smallholder agriculture.

Notes

1   FAO (2009) 'The State of Food and Agriculture 2009', Food and Agriculture Organization of the United Nations, Rome.
2   World Bank (2007) 'World Development Report 2008', p.29. Washington.
3   IFAD (2010) 'Rural Poverty Report', International Fund for Agricultural Development, p.3, Rome.
4   World Bank, op. cit., chapter 1.
5   Oxfam International (2009) 'Investing in Poor People Pays', pp.8–9, Oxfam; Oxfam International (2009) 'Harnessing Agriculture for Development', *Oxfam Research Report*, pp.13–16, Oxfam International.
6   FAO (1995) 'Rural women and the right to food', FAO Women and Population Division, Sustainable Development Department, shows that women in sub-Saharan Africa contribute 60–80 per cent of food for both household consumption and for sale, [Online] http://www.fao.org/docrep/w9990e/w9990e10.htm [accessed 11 March 2011].
7   Women own less than 10 per cent of land resources in many countries, and just 5–15 per cent of agricultural training is targeted at women.
8   For a study area in Burkina Faso, Udry et al. (1995) show a likely 10–20 per cent increase in output if inputs such as labour and fertilizer are reallocated from men's to women's plots in the same household. *Source:* Udry, C., Hoddinot, J., Alderman, H. and Haddad, L. (1995) 'Gender differentials

in farm productivity: implications for household efficiency and agricultural policy', *Food Policy*, 20(5).

9 The Small Farmers, Big Change workshop held in Oxford on 12–14 May 2009 was a major learning opportunity, bringing together 45 participants from 18 countries and three Oxfam affiliates (Oxfam GB, Oxfam America, and Oxfam India). Five external speakers also shared inputs, drawing on their experiences with the UK Department for International Development (DFID) and a variety of research institutions and NGOs.

10 Overall, government spending on agriculture stood at 11 per cent of total government spending in 1980, and had declined to 7 per cent in 2002. Many African countries undertook a commitment to increase agriculture's share of spending to 10 per cent in the 2003 Comprehensive Africa Agriculture Development Programme (CAADP) Maputo declaration. However, by 2005, only six out of 24 governments had met this commitment. Official development assistance (ODA) to agriculture dropped by 75 per cent during the late 1980s and early 1990s. Total donor investments in agriculture have since remained low, at around $4 bn per year. In 2007, US and EU ODA commitments to agriculture increased slightly to $1.2 bn and $1.4 bn respectively, compared with the astonishing $41 bn and $130 bn lavished on their own agriculture sectors in 2006. Oxfam International (2009), op. cit.

11 See e.g. IFAD (2010) *Rural Poverty Report 2011*, IFAD, Rome.

12 More details on this and other tools for promoting women's economic leadership in agricultural markets can be found on Oxfam's online community website: www.growsellthrive.org This is a social networking platform that communities of practice can use to support both online and face-to-face interaction, collaboration and learning.

13 See Oxfam International (2010) 'Think Big. Go Small: Adapting business models to incorporate smallholders into supply chains', *Briefings for Business*, 6, Oxfam International.

14 Women's economic leadership in agricultural markets aims beyond 'more women producers' or 'women as committee members', or improving what women already do, although these are good outcomes. What is new in this approach is an explicit understanding of how agricultural market institutions and services can reinforce gender inequalities in roles and ingrained beliefs about appropriate roles for men and women. Equally, changes in market systems can be a significant lever for longer-term change in gender relations, sparking wider changes at community and household levels. Rather than focus on 'barriers', the starting point for Oxfam is an explicit process to identify market opportunities for women producers to gain new roles and power in agricultural market chains. See www.growsellthrive.org

About the authors

Sally Baden is Senior Global Adviser, Agriculture and Women's Livelihoods, Oxfam GB.

Claire Harvey is Communications Manager (Content Development), Policy and Practice Communications Team, Oxfam GB.

CHAPTER 1

The voice of many – Honduran citizens hold the state to account to secure farm investment

Sonia Cano

Contributors: Maritza Gallardo and Hector Ortega

Rural communities in Western Honduras have long been marginalized, suffering high levels of poverty and inequality. Governments have shown little interest in smallholder agriculture and as a consequence have invested little in the region. Following ten years of action to build civil society, however, communities and local authorities are now working together to claim their rights. Working progressively at different levels – community, municipality, regional, and national – Oxfam and its partners have campaigned to promote small-scale agriculture as a viable means of building livelihoods and reducing poverty. As a result, new funding is beginning to have an impact in rural areas.

Introduction

Honduras suffers from some of the highest poverty levels in Latin America. Of its population of about 7.2 million, 55 per cent of urban households and 70.8 per cent of rural households live below the poverty line, and one-quarter of the population – 1.6 million people – live in conditions of extreme poverty.[1] Although it is classed as a middle-income country, Honduras scores 0.667 on the UNDP's Human Development Index (HDI), significantly below the Latin American average of 0.797,[2] due to its highly unequal income distribution.

Although 70 per cent of the population live in rural areas, over the years the state has reduced funding and technical assistance to the rural sector.[3] Instead, governments have prioritized support to large- and medium-scale commercial agriculture for export crops. This has hampered the country's capacity to produce food for domestic consumption and has increased its dependency on imported food. With rising global food prices and more land being used to produce ethanol for fuel, Honduras is increasingly vulnerable to food insecurity.

Citizens have historically been excluded from decision-making. Between 1963 and 1982 the country was ruled by a succession of military governments.[4] The 1980s saw a culture of authoritarianism and party political sectarianism, with violent persecution of those opposing the government; many leaders of social movements were killed. While elections have been held more regularly in recent times, real participatory democracy is still a challenge. Governments have generally accepted participation that supports the implementation of their own policies, but have been less accommodating of participation by citizens when the aim is to bring about real change.

Nevertheless, in 1990 the Honduran government initiated a process aimed at decentralizing decision-making. The central government retained responsibility for determining policy, but operational functions were devolved to the municipalities[5] and the private sector. A new Law of Municipalities gave them autonomy to approve budgets, collect payments for local services, operate public utilities, and create mechanisms allowing citizens to participate in local democracy. The municipalities were to fund this new model with the income they collected, while an additional 5 per cent of national income was transferred to them from central government.[6]

More additional resources became available in 1999, when Honduras became a beneficiary of the Highly Indebted Poor Countries (HIPC) initiative.[7] In return for debt relief, the country agreed to design a Poverty Reduction Strategy (PRS), setting out how funds would be invested. This provided a policy framework for negotiating new funds with the International Monetary Fund (IMF) and the World Bank. Honduras also joined in the global commitment to work towards the Millennium Development Goals (MDGs), which necessitated stringent public budgeting. The PRS and the MDGs became reference points to determine how state income – whether from grants, new foreign debt contracts, or domestic income – could be used.

As the allocation of resources and operational decision-making became increasingly decentralized and the state developed a greater focus on reducing poverty, so the opportunities for civil society to influence budget decisions at municipal and regional levels increased dramatically. Oxfam's engagement in the process began in 1999, through a programme designed to support the development of a movement of active citizens in Western Honduras.

Empowering citizens in Western Honduras

In the country as a whole, rates of illiteracy range between 50 per cent and 70 per cent, and malnutrition, poor housing conditions, and limited health services mean that diseases such as polio and tuberculosis are rife.[8] Western Honduras, comprising the departments of Lempira, Copán, and Ocotepeque, has experienced particularly acute problems of poverty, exclusion, and marginalization. The six departments of Honduras with the lowest HDI rankings are Lempira, Copán, Ocotepeque, Intibucá, Santa Bárbara, and La Paz, where the population is mostly of indigenous origin (Lenca and Maya Chortis ethnic groups).[9]

Initially the focus of Oxfam's intervention was on creating space for citizens to participate and engage in dialogue with those who governed them, with the aim of finding ways to tackle the root causes of poverty. However, it was also clear that improving the capacity of rural people to generate income was critical in overcoming poverty and exclusion. Oxfam introduced a programme under its Global Agricultural Scale Up Initiative (GASUI), which consisted of two key components. First, the programme aimed to demonstrate the viability of agriculture as a livelihood activity that could link rural people to profitable markets (see Box 1.1); second, it engaged in advocacy and campaigning work that would be critical in changing local, regional, and national policies on agriculture and creating an enabling environment to support smallholder farmers.

The first phase of work to influence policy and investment took place in 1999–2000, and was aimed at building active citizenship in the Western region. In 1998 the Association of Non-Governmental Organizations of Honduras (ASONOG) carried out an assessment on democracy and citizen participation in the region. Oxfam agreed to work in partnership with ASONOG on a process aimed at closely linking citizen participation with human development, in a way that would change the political culture in Honduras.

In that year Hurricane Mitch struck the country with devastating force, leaving 5,000 people dead, most of its crops destroyed, and damage estimated in the billions of dollars.[10] In the aftermath of the disaster, ASONOG supported municipal authorities with training in relevant municipal laws, planning, budget formulation, tax collection, and resource management. It also provided them with funds to develop small projects of social benefit for the poorest and most excluded people, after consulting with the communities themselves about the relevance of each project.

The programme began at the community level. Oxfam and ASONOG set out to engage with citizens and to encourage them to take action to claim their democratic rights. Successful strategies included:

- Using theatre, traditional dance, music and poetry festivals, and radio and TV programmes to educate people about the importance of democracy and about what being a citizen involves.
- Developing 'advocacy schools' for different population groups, offering citizens training in advocacy and campaigning. These schools brought together different groups, leading to a collective dialogue and analysis of the situation in Honduras and the formation of networks.
- A 'training of trainers' model for communicating knowledge. Local leaders who received training committed to pass on what they had learned within their own communities or social organizations.

Combining recreation and education was a very efficient way of overcoming the apathy felt by many citizens towards the democratic process and of promoting an attitude of collaboration. In particular, collaboration on small community projects made it possible to lay the foundations for a more transparent use of municipal resources, as community members were informed about the quality and costs of materials and contracts agreed for machinery and equipment.

Box 1.1 Demonstrating the viability of smallholder agriculture: small farmers and their businesses lead the way

Oxfam has been working with local partner the Christian Organization for the Integrated Development of Honduras (OCDIH) since 2006 to demonstrate the viability of small-scale agriculture as a profitable livelihood option. This means turning small family plots used to grow basic grains into integrated, diverse farms producing a range of fruits, vegetables, and livestock for consumption and for market. The model also aims to demonstrate the most effective and profitable ways of investing in the sector. In particular, the programme has supported:

Improved natural resource management: Through the development of irrigation systems and farmer-to-farmer training programmes, the programme has reached over 20,000 beneficiaries – increasing yields of corn by 54 per cent from 2006 to 2010 and incomes by 27 per cent in the same period. Farmer training programmes and demonstration farms have encouraged farmers to diversify the crops and livestock they produce, as well as to adopt new soil and water management practices that reduce the need for costly inputs, such as fertilisers.

Access to credit: Poor access to credit hampers the ability of farmers to invest in their farming activities. Formal financial institutions offer credit at extremely high interest rates. The programme has established over 90 savings and loan community funds (*cajas rurales*); in 2008 these had accumulated capital of $294,370 and over 2,000 beneficiaries, 48 per cent of whom were women. Lower interest rates for women have improved their access to loans and their ability to repay them, and have also increased their ability to make autonomous decisions about the use of credit.

Linking to new markets: As well as boosting production, the programme has supported farmers to develop agriculture through small businesses in order to access new markets and achieve better prices. It has contributed to increasing producers' sales by $150,000 in the past year in municipal farmers' markets and through wholesalers and supermarket chains. The number of women who have access to goods and agricultural inputs and are active in the market has now reached 31 per cent, far greater than the 1 per cent recorded in 2006. As a result, the baseline of the Economic Justice programme reflects an increase in the contribution of women to household income from 10 per cent in 2007 to 20 per cent in 2008.

This set the stage for the next level of citizen participation, which saw the programme focus its efforts at the municipal level. In response to legislative changes introduced in 2002, which reduced the role of civil society in the country's Departmental Development Commissions,[11] ASONOG worked to support the *mancomunidades* (associations of municipalities). The *mancomunidades* were born out of a joint planning effort between civil society and municipal government, and are made up of representatives of each. They already existed when ASONOG began its civil society participation programme in 1999, but at that time had achieved only low levels of co-ordination and joint management.

Efforts were also focused on the *patronatos* (community councils); these involve community members in decision-making at the local level and raise funds to implement small-scale projects. Associations of community councils (*asociaciones de patronatos*) were set up to create alliances with municipalities in order to raise funds jointly. ASONOG also built up the social network by integrating member organizations (17 NGOs and farmers' associations) into its training and education programme. The programme also built the capacity of partners to mainstream gender, for example by including gender indicators in their planning processes and monitoring the progress of economic empowerment of women in the region. Women's increased influence in decision-making in the use of municipal funds in the Western and Central regions (Intibucá and La Paz) is evident in the increased allocation of special funds for women's projects. Ten per cent of the PRS funds, the equivalent of about USD $8,000, was utilized for projects run exclusively by women.

The programme reached the next level when in 2006 a regional-level co-ordination council – the Western Regional Platform of Honduras, or EROC – was established, consisting of equal numbers of municipal mayors and representatives of civil society.[12] EROC acts as a channel for political discussion with the central government about development issues in the region. It represents a cultural change at the local level, fostering collaboration between actors with different political and party affiliations. The creation of the council was fundamental in enabling the development of a specific PRS for the Western region and in securing additional funding from central government to implement it (see Box 1.2).

The achievement of the Western region in formulating its own PRS is impressive, but at the national level huge challenges remain. Only a small proportion of PRS funding is decentralized, and civil society and the donor community have questioned the government's distribution of funds. It was originally proposed under the PRS to invest 50 per cent of available resources in rural programmes to promote agriculture and micro-businesses, but this has not happened. Instead, successive presidents have applied their own interpretations of PRS principles, making the excuse that small-scale agricultural production is uncompetitive in markets. During the first six years of PRS implementation, there was scarcely any investment in the rural economy, while in 2007–08 it amounted to just 1 per cent of the total received in debt relief.[13]

In response to this, in 2009, the Oxfam-supported programme launched a national advocacy campaign on small-scale agriculture aimed at leveraging change in policy and investment at the national level. Its objectives were the introduction of a reform law for the agricultural sector and a 10 per cent increase in resources for small-scale agriculture in the national budget.

Box 1.2 A critical moment: the development of a regional Poverty Reduction Strategy (PRS)

Honduras's first national PRS was ratified by the government and the IMF in 2001, even though the proposed draft neglected many of the concerns voiced by civil society. In Western Honduras, dissatisfaction with this national plan was channelled into the positive process of creating a regional PRS of its own. Led by EROC, the design of the Western Region Strategy began at the community level and involved thousands of individuals from community associations, women's groups, and municipal authorities. In 2006, the initiative was launched as an alternative to the national PRS.

There are key differences between the two. For example, while the national PRS proposes extractive industries (mining and timber exploitation) as key areas for economic growth, civil society in the Western region (represented by a coalition of 120 national organizations led by ASONOG and FOSDEH, the Social Forum on External Debt and Development[14]) proposes as key actions environmental protection and recovery, the conservation of water resources, and reforestation. While the national PRS proposes tackling food insecurity by increasing food imports, civil society believes that the emphasis should be on access to credit to enable farmers to improve production for domestic consumption. The official PRS proposes distributing debt relief funding according to size of population (which would favour large cities), while civil society in the Western region proposes distribution based on poverty indicators.

The central government originally refused to recognise the validity of the Western region's PRS, but two important factors forced it to change its mind. First was the fact that municipal authorities had been strongly involved in the design of the strategy and were publicly committed to it. Second was the national-level lobbying carried out by civil society in partnership with other networks such as the CCERP (Consultative Council for Poverty Reduction), with the support of the international development community and the World Bank, which publicly recognized the quality and validity of the Western Region Strategy.

ASONOG is leading the campaign jointly with two of its member organizations, the Christian Organization for Integrated Development in Honduras (OCDIH) and the Organization for the Development of Corquín (ODECO).[15] The campaign is active in six regions of Honduras, and at national level is linked to the Meso-American campaign and regionally to Oxfam International's Economic Justice Campaign within Guatemala, Mexico, and Honduras.

The NGOs have gathered evidence on how small producers can be competitive in markets, generating income to pay taxes and to raise funds that contribute to the development of their communities. In 2008 four important groups joined the campaign, giving it a genuinely national reach: the Central de Cooperativas Cafetaleras de Honduras (National Coffee Co-operatives Centre), the Co-ordination Council for Small Farmers' Federations (COCOCH), and local NGOs Asociación Ecológica de San Marcos de Ocotepeque (AESMO) and Coordinadora de Mujeres Campesinas de La Paz (COMUCAP), a rural women's organization.

The campaign was launched nationally in March 2009, with mass events held simultaneously in six regions of the country bringing together small producers, academics, and senior government officials. This level of mobilization was made possible by the organizational work and capacity-building that had started ten years earlier. It has paid dividends: recognizing the social force behind the coalition leading the campaign, the government has agreed to discuss a potential 10 per cent increase in the budget for small-scale agriculture. Additionally, the campaign was featured in the national media for five weeks after it was launched, which helped to generate favourable public opinion.

Results and achievements

Programming efforts over a period of nine years have led to significant changes in the political dynamics of the Western region of Honduras and its ability to access funds from government and other sources, such as the Swedish International Development Agency (Sida) and other NGO and government donors. The participatory approach taken by ASONOG has proved adaptable and has grown in scale over time. Starting in 1999 with ten municipalities, ASONOG is now implementing the intervention with its two partner organizations, OCDIH and ODECO, in 67 municipalities. As a result of this work, civil society has been radically changed. At the municipal level, there are currently more than 400 *patronatos* able to act independently of party political interests, 13 *patronato* associations, five youth networks, 25 women's networks, and 67 citizen transparency commissions which audit the use of funds by the municipalities. The intervention has also led to the creation of EROC, which co-ordinates the efforts of community groups and municipal authorities at regional level.

The programme has improved the region's ability to secure debt relief funding. In 2006, a total of $37 m of PRS funding was devolved to municipalities nationwide, but with conditions attached that not all municipalities could meet.[16] The Western region was the first – and so far the only – region to draw up its own regional PRS, in a joint effort by civil society and municipal leaders. In 2006, it received a total of $8.84 m for projects, the highest amount allocated to any of the country's five regions. Elsewhere, municipalities did not base their budgets on PRS funding and civil society was unable to influence the destination of funds.

As a consequence, Western Honduras has gained greater influence with the national government. Previously, the region was the most isolated and marginalized in the country, starved of funds for public investment and with no civil society representation at national level. The government engaged in dialogue only with politicians and big business. Now, however, the region has a voice and is able to present proposals to decision-makers, who are prepared to listen. EROC has engaged with the Consultative Council for Poverty Reduction (the body set up in 2002 to oversee national PRS funding, representing civil society, donors, and the state) and with the technical unit of the government responsible for analysing projects and municipal investment plans, which have gone to the Secretary of Finance, parliamentary whips, and the President of the Republic himself.

Box 1.3 Defending smallholders' interests at the national level

The capacity of civil society in the Western region to influence national policy was put to the test when the government of Manuel Zelaya, elected in 2006, rescinded an agreement signed by the previous government, led by Ricardo Maduro, to direct 100 per cent of debt relief funds to reduce poverty in the country's poorest municipalities.

Instead, President Zelaya issued a decree ordering the distribution of funds according to population density, and redirected Lempira 900 m ($47.6 m) to finance current expenses and election campaign commitments, leaving just Lempira 700 m ($37 m) to invest in poverty reduction. The decree also reduced the role of the Consultative Council for Poverty Reduction by transferring responsibility for allocation of funds to the National Congress.

However, both provisions were revoked following joint action by civil society in the Western Region, led by EROC, and national-level lobbying by ASONOG and FOSDEH.

This unprecedented improvement in its status has put the region in a much better position to negotiate, and in 2006 it obtained funds of $8.84 m from debt relief and donors to invest in projects under its own PRS. By working at different levels, it has also proved possible to leverage resources from other donors. While negotiating on development issues with regional and national governments, for instance, EROC has established links with the wider donor community. In 2008 it obtained $4.76 m in funding from the Swedish government to support projects in 30 municipalities identified as being the poorest of the 67 in the region. The funds will be managed by the municipalities and audited by the citizens' transparency commissions. Of these funds, 31.7 per cent was used for infrastructure projects, which improved access to agricultural markets, and 4 per cent was invested directly into home-based agriculture, such as growing vegetables and rearing pigs or poultry.

Meanwhile, community associations have raised funds from central government and from donors to develop projects that they themselves have identified as priorities.[17] Between 2006 and 2008, with assistance from OCDIH, community council associations in the north of Copan and the municipality of Lepaera in Lempira carried out 16 fundraising missions, obtaining $1.9 m to develop 83 rural community projects. Of this, 15 per cent went to fund community banks (*cajas rurales*), agricultural production, and rural infrastructure projects to improve transport to markets.

The availability of increased resources has had a clear impact on the livelihoods of smallholder farmers. The funding earmarked for improving infrastructure is supporting work done under the Honduras Agricultural Scale Up programme, improving access to markets for farmers in remote areas. If the national agriculture campaign builds on its promising start and manages to secure a 10 per cent increase in the national agricultural budget, this will have an even more significant impact on the lives of over five million rural Hondurans.

Scaling up and replication

Oxfam and its partners pursued a strategy of scaling up, which relied on their ability to secure financial resources from other actors at community, regional, and national levels. Establishing and strengthening community, municipal, and regional institutions for planning work and transparently managing funds have put in place sustainable structures to ensure the appropriateness and effectiveness of future investment.

Building links between the advocacy component of the programme and work that demonstrates the competitiveness of small-scale producers in agricultural markets has been a critical aspect of this initiative. It has enabled the programme to secure investments far greater than any that a single civil society organization could have invested by itself.

The programme has also been strategic in using its positive experiences at regional level to leverage change at national level. The development of the national agricultural campaign builds on work with farmers to support production and market access as well as capitalizing on the growing influence of the Western region in national government. These experiences have made it possible for the campaign to provide both the government and the general public with real examples of how support to small-scale agriculture can contribute to reducing poverty and how such interventions can be managed successfully. The programme also plans to make the most of opportunities presented by regional and national debates on food security. Oxfam and its partners will strongly advocate for investment in smallholder agriculture as an effective means of providing affordable food for both urban and rural populations.

Another way of scaling up the approach has been to promote replication of the model in other regions. In 2005, using the learning gained in the Western region, development organizations Trócaire of Ireland, DanChurchAid of Denmark, Diakonia of Sweden, and Development and Peace of Canada began a process to develop regional Poverty Reduction Strategies in three departments in the north of Honduras (Atlantida, Yoro, and Colon) and in the Sula Valley in the south of the country.

However, their success was more limited as they were not able to invest in the long-term, bottom-up process of building civil society at the community level. They were therefore not able to address the allegiances of *patronatos* to political parties, and ideological differences hindered them from establishing social alliances and networks. This meant that differing levels of engagement were achieved between municipalities and civil society in each of the regions. With a weaker base, civil society has less power to negotiate with the state. This only serves to emphasize the value of the process of building active citizenship that took place in Western Honduras.

Successes and challenges

A number of factors have contributed to the success of the programme in Western Honduras. Elements that have worked well include the following:

- The initial investment made in training and education on advocacy, democracy, and citizens' rights was minimal compared with the impact and reach achieved by means of participants passing on what they had learned within their communities and through farmers' and women's organizations. This made it possible to create a mass movement by disseminating skills and knowledge throughout the population.
- The strategy of gradually scaling up the intervention – first through action at community and municipal levels, then at the regional level, until finally reaching the national level – proved effective. It allowed an incremental build-up of human capital, stimulating networking amongst civil society and a systematic weaving of alliances with municipal authorities.
- The capacities developed by citizens and the alliances established with local governments gave the campaign legitimacy in the eyes of the government and of donors. This facilitated fundraising activities and helped to progress the decentralizing of resources in favour of the poorest rural communities.
- The creation of a strategic alliance between NGOs, such as ASONOG, OCDIH, ODECO, and others, and farmers' and women's organizations to promote the agriculture campaign was a crucial factor. A high level of inter-agency co-ordination allowed the NGOs to pool their expertise and learning in a spirit of co-operation. Their proposals were seen by the government not simply as another demand from the farming sector, but as the reasoned and articulate response of an alliance of social actors acting in pursuit of a common goal.

Some important challenges remain, however. Among them are illiteracy and low levels of education, and actions taken by political parties and charitable organizations that hinder social empowerment of citizens – for example, badly managed donations that encourage passive dependency amongst communities and bypass institutions established to manage these resources.

However, the most significant threat to the programme is the political crisis in Honduras. President Zelaya was sent into exile on 28 June 2009 amid a power struggle over his plans for constitutional change. Prior to the coup, most space in the media was taken up by the activities of the main political parties, which hindered the programme's advocacy and campaigning, as there was little opportunity to position other issues on the national agenda. Following the coup, there was a break in relations between central and local governments, over lack of resources and political issues. What little aid there is has become politicized and centralized.

Following a period of interim government, elections were held in November 2009. Despite international criticism of the coup, President Porfirio Lobo was sworn into office in January 2010 as Zelaya went into exile. Following the coup, the programme strategy has been redefined to respond to the changing context. A new emphasis has been put on developing the leadership capacity of youth and women's groups, to further enable them to develop their own agendas and their ability to take part in political dialogue and to contribute to the construction of new state law. However, continued insecurity and disregard for human rights have resulted in threats, violence, and political persecution of leaders, so activities promoting protection have become more key to the programme.

Conclusion

The most significant conclusion to be drawn from this experience is the importance of the bottom-up process in influencing policy and investment decisions. It is clear that the work done to mobilize civil society at different levels was central to the success of the national campaign, especially as the many efforts previously undertaken in this area had failed. The impact of the work can be seen in the relationships established between communities and municipalities and between municipalities and the national government. As a result of this engagement, the state is much better equipped to allocate funding to activities that are relevant and desired by community members themselves.

The ultimate outcome of the national campaign cannot yet be known, but its immediate achievements, in terms of the media profile given to the launch and the willingness of the government to engage with its proposals, could never have been imagined just ten years ago.

Notes

1   UNDP (2006) *Informe sobre Desarrollo Humano Honduras 2006* (Human Development Report, Honduras 2006) p.40, UNDP, Honduras.
2   Ibid., p.27.
3   Ibid., p.67.
4   The last Honduran military dictatorship was led by Policarpo Paz García, who ruled from 1978 to 1982.

5 The 18 departments of Honduras are further divided into 297 municipalities. The municipalities have elected mayors and municipal councils rather than departmental governors who are appointed by the President. Municipalities vary considerably in size, and their governments are constituted accordingly. Those with a population of under 5,000 have four council members, those with 10,000 have six, those with more than 10,000 have eight and a mayor, and those with more than 80,000 have ten and a mayor. See US Library of Congress: http://www.countrystudies.us/honduras/88.htm [accessed 15 March 2011].

6 UNDP (2006) op. cit., pp.83–7.

7 This was precipitated by Hurricane Mitch, which devastated the country in 1998. The IMF had repeatedly rejected Honduras's inclusion in the HIPC initiative, arguing that its economy was healthy. After Hurricane Mitch, however, this argument was untenable and Honduras was granted debt relief (Foro Social de Deuda Externa y Desarrollo de Honduras – FOSDEH).

8 FAO (2005) 'Status of Food Security in Honduras', p.35, [Online] http://www.fivims.org/index.php?option=com_content&task=view&id=107&Itemid=104 [accessed 9 April 2011]

9 Around 90 per cent of the country's population are *mestizo* (of mixed indigenous/European descent), with 2 per cent of African origin and 1 per cent European (US Library of Congress). According to the Confederation of Autochthonous Peoples of Honduras (CONPAH), 7 per cent of the population is made up of indigenous ethnic groups, among them the Lenca, Chorti, Garífuna, Tolupan, Pech, Nahoa, Miskito, and Tahuaca.

10 World Bank, 'Honduras country brief', [Online] http://web.worldbank.org/WBSITE/EXTERNAL/COUNTRIES/LACEXT/HONDURASEXTN/0,,contentMDK:21035522~pagePK:141137~piPK:141127~theSitePK:295071,00.html [accessed 15 March 2011].

11 President Ricardo Maduro reformed the Decree for the Creation of the Departmental Development Commissions (CODEP) in mid-2002. This reform reduced the participation of civil society in decision-making, with increased responsibility given to Congress and the Governor Co-ordinator, both political positions designated by the President.

12 In Spanish, EROC stands for Espacio Regional de Occidente.

13 According to government figures, in 2000–02 Honduras received $81.7 m in debt relief and spent $67.2 m; in 2003–05 it received $718.2 m and spent $14.5 m; and in 2006–08 it received $215.1 m and spent $141.3 m. The main areas of expenditure were salaries for teachers and police, medicines, and social protection and infrastructure. *Source:* World Bank, 'World Bank Strategy for Honduras', [Online] http://web.worldbank.org/WBSITE/EXTERNAL/COUNTRIES/LACEXT/HONDURASEXTN/0,,menuPK:295083~pagePK:141132~piPK:141105~theSitePK:295071,00.html [accessed 15 March 2010].

14 In Spanish, Foro Social de la Deuda Externa y Desarollo de Honduras. FOSDEH is a national NGO network created in 1996 by an ASONOG initiative.

15 OCDIH (Organismo Cristiano de Desarrollo Integral de Honduras) and ODECO (Organización para el Desarrollo de Corquín) are both members of the ASONOG federation.

16 At the same time, the government withheld $104 m to implement elec-
   tion promises and to support charitable causes promoted by the First Lady.
17 Members of community associations have been trained as fundraisers and
   supported by OCDIH.

About the author

Sonia Cano is the Programme Co-ordinator, Honduras, Oxfam GB.

CHAPTER 2

Strength in numbers – fishing communities in India assert their traditional rights over livelihoods resources

Mirza Firoz Beg

Villagers in the Tikamgarh and Chattarpur districts of Madhya Pradesh traditionally had the right to fish the region's ponds, but had lost control of these valuable resources to landlords and contractors. Despite encountering violent opposition, the fishers began organizing to reclaim control of the ponds. They established village co-operatives and formed a federation that gave a strong voice to the region's fishing communities. By 2008, fisher co-operatives controlled 151 ponds, with nine run by women's groups. In 2008, their campaign persuaded the state government to revise its fisheries policy, introducing a new law that protects the rights of traditional fishing communities and contains provisions that should help to improve livelihoods in the drought-hit region.

Introduction

The Bundelkhand region of central India is a semi-arid plateau that encompasses six districts of northern Madhya Pradesh and seven districts of southern Uttar Pradesh. Much of the region suffers from acute ecological degradation due to topsoil erosion and deforestation, making the land unproductive. It is also one of the least developed regions in the country in terms of per capita income and literacy levels. In Madhya Pradesh the proportion of the population living below the poverty line was 38.3 per cent in 2004–05,[1] while in 2008 the male adult literacy rate (age 15 and over) was 73 per cent and the female adult literacy rate was just 48 per cent.[2]

Fishing is an important economic sector in India. Inland and marine fisheries together account for around 1 per cent of national GDP and make an important contribution to foreign exchange earnings; in 2004 these amounted to $1.36 bn.[3] FAO estimates that the sector supports around 14.7 million people, with two million directly employed in full-time or part-time fishing activities and nearly four million engaged in ancillary activities such as net-making and fish vending.[4] The state of Madhya Pradesh is a significant contributor to India's growing inland fisheries sector, with over 335,000 hectares of ponds and reservoirs.[5] The state's annual production of fish amounts to 61,500 tonnes, and it has approximately 1,680 fisher co-operatives with 58,500 members.[6]

Despite the industry's potential, traditional fishing communities in Madhya Pradesh have struggled to sustain their livelihoods. In 1998, when Oxfam began working in Bundelkhand, the rights of traditional fisherfolk had been largely eroded and contractors, given access by village landlords, had gained control of the ponds. Many of the region's 45,000 fishing families had become waged labourers, badly treated and poorly paid, and at risk of losing access to the resources that provided them with a livelihood. The state government provided little or no support for fishing communities, and local officials paid no attention to their grievances.

With the support of Oxfam, fishing communities in the two Bundelkhand districts of Tikamgarh and Chattarpur began fighting to re-establish their rights over the fishing ponds in the late 1990s. VIKALP, a local NGO specializing in sustainable development, also played a crucial role in helping communities to set up village co-operatives to fish the ponds. Their struggle gained momentum as fishermen and women from other villages joined their campaign. This led eventually to local fishing communities regaining control over 151 ponds for the cultivation of fish (and of crops in the dry season) – a huge achievement. Furthermore, the community-led campaign persuaded the state government to introduce a new fisheries law that guarantees rights to the ponds, improves access conditions, and promises to enhance livelihoods for many more people in the region. With this support, the groups are improving both the ponds and their own levels of organization, in the process emphasizing the empowerment of women.[7]

Marginalized fisherfolk fight back

The Dhimar caste of fisherfolk have traditionally enjoyed fishing rights on ponds and reservoirs (known locally as 'tanks') in the Bundelkhand region. Many of these ponds were built as long as 1,000 years ago. For many years, however, they were neglected and local people fished only for their own consumption.

In 1967 responsibility for leasing the ponds became the responsibility of the *panchayats* (village governing bodies). While on paper fisher co-operatives were meant to have priority, over the years that followed upper-caste landlords, middlemen, and contractors realized the market potential of fish and used their influence in the *panchayats* to take control of the ponds. The terms of the leases they secured were unfair, and local fishers became marginalized, forced to provide low-paid labour for the landlords and contractors. Any attempt to fish on their own account was treated as theft, and fishers were subject to abuse and even to physical violence.

In 1995, a group of 12 young Dhimar fishers in Madiya village took organized action to reclaim their traditional fishing rights. They negotiated with the *panchayat* and district authorities to take over the lease of the Achhrumata pond, despite opposition from the contractor who had previously held it. The fishers stocked the pond with fish, but then at harvest time were confronted by thugs hired by the contractor. A pitched battle ensued, which ended with the fishers' huts being burned down.

Undeterred, the Achhrumata fishers persuaded the local police to accept an official complaint, which was seen as an important symbolic blow against the landlords. This unprecedented success, together with legislation passed in 1996, sparked a wave of organization in poor fishing communities, with other fishers filing formal complaints about attacks made on them.

VIKALP's mission is to work with marginalized and resource-poor communities, particularly women, children, and youth, and to empower them with knowledge and skills, so that communities can achieve development and ensure food security themselves through sustainable practices of natural resource management.[8] With VIKALP's help, this first group of fishers organized campaigns to encourage others to set up co-operatives and, as their self-confidence grew, fishers took over several other ponds. By the end of 1998, co-operatives had control of ponds in Mandiya, Kakuani, and Daretha. Despite still having to fight the landlords, the experiences of these pioneering groups demonstrated the returns available if fishers were able to organize and control the ponds. However, it was also clear that changes in policy would make it much easier for fishers to retain control in the long term.

In 1996 the fishers' protests led the Fisheries Minister in the Congress Party Government of Madhya Pradesh to push through 'fish to the fisher' legislation that granted leases to fisher co-operatives and gave *panchayats* a clearer framework for managing leases. Although contractors used 'dummy' co-operatives and other tricks to get around the legislation and retain their control of ponds, the new law prompted a wave of co-operation and organization among fisher communities.

Box 2.1 Reaping the rewards of the struggle

In the village of Dumduma a group of fishers took the Ganga Sagar tank on lease, despite demands for compensation by contractors for the investments they claimed to have made. The fishers negotiated the Rs. 50,000 ($1,083) demanded down to Rs. 32,000 ($693) and with great difficulty managed to scrape this sum together, along with an additional Rs. 12,000 ($260) to stock the tank with fish fry. Their first-year returns on the fish catch amounted to Rs. 300,000 ($6,500), despite the contractors extracting a further Rs. 32,000 ($693) from the group.

In 1998, Oxfam began supporting the fisherfolks' efforts. A meeting held in Kakoni attracted around 50 fishers, together with representatives of VIKALP and Oxfam, who agreed to help mobilize financial support. Representatives from co-operatives began meeting monthly, and solidarity between the different groups grew. Unsettled by the growing level of fisher organization, the landlords and contractors attempted to prevent them from meeting, making threats and attempting to create splits between co-operatives. There were many instances of nets being stolen, beatings, and abuse.

By 2000, fisher co-operatives controlled 22 ponds – though the majority remained beyond their reach. Challenges included the high cost of leases; lack of capital investment, quality fish stocks, and market linkages; and restricted rights on the use of pond beds to cultivate water lotus (*kamalgatta*) or other crops during the dry months. Even where fishers did control ponds, they often had to pay up to half of their returns to contractors to cover existing debts or to compensate them for investments in fish stocks.

Continued campaigning spread the word to a further 56 villages, and in 2002 a mass meeting of 10,000 fisherfolk was held in Tikamgarh, followed by the presentation of a 14-point charter of demands to the district collector (the most senior government official in the district), who agreed to take up the matter with the state government. In May 2002 a further mass meeting of fisher representatives was organized with the collector. This resulted in the district administration taking stronger action to remove illegal contractors and to punish co-operative officials illegally colluding with contractors.

Box 2.2 Women gain in self-belief

Women in the fishing communities began organizing self-help groups in the late 1990s. By 2001, many women's groups had accumulated savings of Rs.15,000–20,000 ($330–440) and started making loans to men's co-operatives for the purchase of fish fry and other inputs.

A women's group first took control of a pond in Birora Kheth village in Tikamgarh in 2002. The women helped to dig the new pond, and when it was complete, put forward their claim to the local *panchayat*. The district collector agreed that the pond would be allocated to whoever could prove themselves capable of fishing it – which the women duly did. They now hold the lease on the pond, manage it, and employ the men to do the fishing work.

This was followed by another women's group taking over a pond in Daretha. Now women's groups have control over nine ponds in Tikamgarh and five in Chattarpur, stocking them with fish and harvesting and marketing their catches themselves.

VIKALP, with support from Oxfam, helped facilitate this process, with a focus on building capacity and empowering the community, especially women. A central part of this process was the formation of a community-based organization, the Achhrumata Machhuwara Sangathan, an umbrella group representing the various fisher co-operatives and women's self-help groups. Set up in 2002, the Sangathan's objective was to work with co-operative members to improve the running of the groups and to help them increase their returns, as well as to encourage the formation of further co-operatives (see Box 2.3).

Box 2.3 Giving the community a voice

The concept of a community body to represent fishers and co-ordinate their campaign for change was put forward in 2002. The Achhrumata Machhuwara Sangathan is made up of representatives from all the registered co-operatives and unregistered self-help groups that have taken over fishing rights to ponds.

A management committee was elected in 2005, consisting of nine representatives from fishing communities (including three women) and one representative apiece from VIKALP and Oxfam. Elections in 2007 saw the number of women committee members increase to four.

The organization provides a number of services for fisher groups, including:

- A seed fund (Rs. 256,000 ($5,600) by 2008) that offers loans to member groups;
- Inputs, such as fish spawn, at reduced rates;
- Boats and nets that can be rented at harvest times;
- Two fish collection centres that reduce transport costs and enable fisher groups to market their fish more easily;
- Technical support on lease management, fish production, and organizational development (e.g. book-keeping), and help in accessing government schemes such as insurance.

Most importantly, the Sangathan helps give fishing communities a voice, where previously they had none. Previously, government officials asked for bribes and officials and politicians never listened to fishers, according to Satish Raikwar, president of the Garkhuan village co-operative in Chattarpur. However, he says, 'Today, because the organization is so strong and more and more people are joining, we can put pressure on the government and political leaders.'

With their new-found confidence, the fisherfolk also began to emerge as a political force and became better represented in the *panchayat* structures that control the pond leases. In 2006, seven of the 15 fisherfolk who stood for election won *panchayat* seats.

Progress towards a new fisheries policy

The fishers' campaign for a new policy initiative on inland fisheries in Madhya Pradesh took an important step forward in April 2007, with a state-level workshop organized by VIKALP. Following this workshop, a policy revision committee was formed, consisting of community members and representatives of VIKALP and of another fisherfolk federation, Tawa Matasya Sangh; this committee drafted a policy outline, which it presented to the Department of Fisheries. A mass meeting held in Chattarpur in November that year was attended by an estimated 3,000–4,000 fishers, with the Madhya Pradesh Minister of Fisheries, Shri Moti Kashyap, as chief guest. In response to the fishers' concerns, the minister promised to formulate a new fisheries policy that would give priority to traditional fisher castes, peg the annual cost of leases, facilitate pond bed cultivation, and introduce insurance and other schemes for fishers.

The minister held further meetings with representatives of the co-operatives, and another mass meeting took place in Jabalpur, where the Chief Minister of Madhya Pradesh, Shri Shiv Raj Singh Chavan, outlined the main points of the new policy. Finally, the long-awaited Madhya Pradesh Inland Fisheries Policy was introduced in August 2008 (see Box 2.4). The new law has many positive features, including a system of decentralized governance through the *panchayat* system and a well-defined support and extension role for the Department of Fisheries. Of particular benefit to fishers are the new rules on leases, including longer lease periods, lower costs, and a single lease for fishing and water-based crops.

Helping to make change happen

Having started in 1998, Oxfam's Bundelkhand programme became part of the NGO's Global Agricultural Scale Up Initiative in 2007. Both Oxfam and VIKALP adopted a more explicit focus on supporting the capacity of fisherfolk to develop sustainable business enterprises and to enhance their livelihoods through a 'power in markets' approach that increased their ability to access and effectively participate in markets. Adopting this way of working did not diminish the importance of the work to support fisherfolk to secure rights over their ponds or the efforts to change policy. Rather, these were essential elements in creating an enabling environment within which fisherfolk could develop profitable livelihoods.

By the end of 2008, the programme had directly benefited an estimated 12,000 households, allowing them access and rights to use 151 ponds, nine of which were controlled by women's groups. For the future, there are plans to reach 50,000 families by 2012 and 100,000 by 2016.[9] It is anticipated that the fisheries policy approved in 2008 could potentially benefit millions of fisherfolk across Madhya Pradesh. After more than 10 years of activities, an analysis highlights the following factors behind the programme's success in achieving change at scale:

Box 2.4 The Madhya Pradesh Inland Fisheries Policy 2008: main points

- Priority in fish cultivation is given to traditional fishing castes. Individuals living below the poverty line have priority for ponds up to one hectare in size.
- The lease period has been increased from seven years to ten years (the longest of any state in India).
- To determine the cost of leases, ponds are divided into five categories depending on size, rather than two – thus making smaller ponds cheaper. The annual lease amount remains fixed for the duration of the 10-year period (compared with a 10 per cent annual increase previously).
- Leases are valid for all uses, including the cultivation of crops such as water lotus and water chestnut in pond beds during dry periods.
- Revenues collected from leases are held by the *panchayats* and used for fisher welfare activities.
- Co-operative societies should have a minimum of 20 members, more for larger ponds.
- Women should make up at least 33 per cent of the membership of any new co-ops registered.
- In the case of drought or natural disaster causing loss of fish stocks or damage to pond structures, the lease amount for that year will be waived and the government may provide some financial compensation.
- Ponds should be managed sustainably – annual fishing bans will be enforced, along with restrictions on draining water from ponds.
- Identity cards and self-employed credit cards are to be introduced for fishers, to enable them to access loan facilities more easily.
- Two representatives from fishing communities will be involved in future reviews of the fisheries policy.

Selecting the right issue

Selecting policy change as a key activity meant that the programme was able to leverage wide-reaching change from a small initial investment. In addition, it was important to identify a policy issue where there was already a community-led initiative with its own momentum and energy. This meant that Oxfam was able to amplify the efforts of the fisher communities and maximize the impact of the work already being done.

Choosing the right interventions to strengthen the movement

The ability to identify gaps in the movement and to provide strategic support to fill them was essential to the programme's success. In particular, programme staff have pointed towards the investment in strengthening the governance structure and organizational capacity of the Sangathan as one of the most effective interventions. In addition, the convening role played by Oxfam and VIKALP was extremely important in attracting relevant actors and experts to support the initiative and in strengthening women's participation and leadership in the community organizations. Commissioning high-quality research on the situation and engaging with the media were also critical in convincing the government to act.

A strategic approach was taken to leveraging support from existing government schemes. For example, labour has been mobilized under the National Rural Employment Guarantee Act (NREGA), which offers rural people 100 days of paid employment per year on public works schemes.[10] In the Bundelkhand region, pressure on the government to implement this act has provided fisherfolk with valuable paid employment and has improved pond infrastructure. In 2008–09, the government committed to renovate 449 ponds under this scheme. In total, more than $290,000 has been mobilized though government labour schemes.

Changing context, changing approach

Oxfam and VIKALP worked with fishing communities over a 10-year period; adapting the focus of the programme in response to changing circumstances was an important part of the model. Initially the programme concentrated on supporting fisherfolk in their struggle to gain control of the ponds. This meant that community organization and district-level campaigns were prioritized. However, over time, engagement with state-level policy became a more pressing issue and VIKALP focused much more on advocacy and on facilitating the engagement of key stakeholders in the policy consultations.

Supporting diversified livelihoods

The programme focused on enabling fisherfolk to capture greater benefits in the fish market – by improving their ability to organize and market their produce jointly. The development of fish nurseries managed by women has provided an additional source of income, as well as improving supplies of inputs. Management of nurseries by women's groups has given women an opportunity to assume a leadership role and to gain recognition and acceptance as managers in the community. However, as ponds are increasingly frequently affected by drought, the cultivation and collective sale of vegetables in pond beds during dry periods to provide an alternative source of income has also been encouraged. In 2006 a total of 227 hectares of dried pond beds were used for vegetable cultivation, which generated income of $36,937 for local communities. The importance of this dual usage of the ponds was acknowledged in the policy framework, and the new lease structure reflects this important development.

Box 2.5 Hatchery enterprise spawns growth in profits

Over the past 20 years, the introduction of new varieties of fish and the practice of stocking ponds with fish fry raised in hatcheries have greatly increased
fish yields. A pilot project in Tikamgarh to raise fish fry is proving a commercial
success.

In the village of Madiya, villagers have set up a hatchery to raise high-quality
fish fry, which they sell directly to other co-ops to restock their ponds. The
hatchery does not require much water, so this is a sustainable activity even
when water levels are low. It is also very profitable, and provides valuable extra
income. Other co-ops in the area are now looking to replicate the initiative.

This new enterprise, which generates income while providing a valuable market
service, is part of the support the programme provides to fishing co-operatives.
In addition, it has helped co-operatives to manage their production more
effectively, obtain price information from different sellers, and access markets
offering the best prices. Being organized also enables the fishers to sell their
produce in bulk, which reduces transport costs and enables them to bargain
for a better price.

Source: Oxfam (2008) *Netting Profits*, DVD.

In Bhangwa village in Bundelkhand, the fishing community has developed
its own hatchery and no longer has to buy fish fry from government or private
contractors. According to Bhuminia, a member of the community, this can save
a significant amount of money. 'It cost us Rs. 40,000 ($880) to develop the
hatchery, and each one of the co-operative members contributed Rs. 1,000 for
this. We have grown 10 lakh (10,000,000) fishlings from this hatchery, and
saved ourselves Rs. 70,000 ($1540). This is good for our village.'

Scaling up the initiative

The Bundelkhand Fisheries Forum has been established to advocate for
the implementation of fisheries policy in the region. The project is being
expanded to include two further areas in the Bundelkhand region, Damoh
and Sagar. In the existing areas, Chattarpur and Tikamgarh, the number of
ponds is to be increased from 150 to 600. The project is also scaling up by
helping fisherfolk to move up the value chain. From working on small and
medium ponds, communities will now take on the challenge of asserting
their rights over large ponds, which will not only strengthen the movement,
but will also help them gain market power through control over higher
volumes of fish.

Challenges faced

Policy alone is not enough

Although the 2008 fisheries policy has brought significant advantages for fisher communities, there is still a need to address some limitations and ensure that the policy is successfully implemented. For example, the new law lacks provision for increasing supplies of quality fish fry and expanding market linkages, and does not address environmental issues such as pollution control, especially in rivers. Although the roles and responsibilities of some actors (such as the Department of Fisheries) are clearly defined, those of others (such as government departments responsible for irrigation, agriculture, and revenue) are not. These are all areas where Oxfam's programme provides a model that demonstrates the possibilities of the initiative to government departments.

Involving women

Although the 2008 policy stipulates that new co-operatives should have a minimum of 33 per cent women members, it does not set out any strategy to increase women's participation. In practice, there are many social and cultural barriers that prevent the adequate involvement of women in the Sangathan. One positive, though unintended, impact in this area has been the emergence of a women's organization addressing issues of violence. However, there is scope to invest more in the entrepreneurial and leadership capacities of women's groups if the programme's gender objective is to be achieved.

Key lessons learned

The development of the new fisheries policy in Madhya Pradesh was a process that began with proactive community mobilization and was based on a consultative approach, facilitated by civil society organizations, and involving the active participation of government officials and political leaders. Key lessons learned from the process include the following:

- Securing rights over assets is a critical aspect of the power in markets approach, but in order to transform people's livelihoods, policy change around asset rights must be complemented by the development of producer-owned enterprises and must work to improve access to both markets and the services that support them (credit, inputs, training, transport, etc.).
- Movements for policy change need to be driven and owned by the community, but are likely to require external support in developing a solid evidence base and in strengthening the governance of community organizations.

- Work is required to ensure the participation of women, where social and cultural barriers make this challenging. However, improvements in economic organization (i.e. through the development of savings groups) offer well-proven opportunities for women to gain power in markets.
- Long-term engagement leads to better and more sustainable results, but strategies need to adapt and change over time in response to changing circumstances.
- Practical limitations that could affect implementation need to be identified and addressed appropriately by the government.
- The example of Madhya Pradesh provides opportunities for other states to review their inland fisheries policies and to make changes that will benefit fishing communities.

Notes

1 Planning Commission, Government of India.
2 World Bank, 'India at a Glance', latest year available 2002–08, available from: http://devdata.worldbank.org/AAG/ind_aag.pdf [accessed 15 March 2011].
3 The sector contributed 1.07 per cent of total GDP in 2003–04. FAO (2009) Fishery and Aquaculture Country Profiles, [Online] http://www.fao.org/fishery/countrysector/FI-CP_IN/en [accessed 15 March 2011].
4 Ibid.
5 Muralidharan, C.M and Mishra, N. (2009) 'Community-led inland fisheries policy development – Madhya Pradesh sets the trend', paper prepared for Oxfam's Small Farmers, Big Change event in May 2009.
6 Madhya Pradesh Department of Fisheries.
7 Green, D. (2008) *From Poverty to Power*, pp. 146–7, Oxfam International, Oxford.
8 See http://www/VIKALPA.co.in/index.html [accessed 15 March 2011].
9 King, R. (draft), revised by S. Mahajan, N. Mishra, J. Zaremba, and S. Baden (September 2008)'Oxfam case study of Livelihoods Programme in India'.
10 NREGA is designed to enhance the livelihood security of households in rural areas. It provides at least 100 days of guaranteed waged employment per financial year to every household whose adult members volunteer to do unskilled manual work at the minimum wage of Rs. 100 ($2.15) per day.

About the author

Mirza Firoz Beg is Programme Officer, Oxfam India.

CHAPTER 3

Leading by example – how cities came to link rural producers with urban food markets in Colombia

Aida Pesquera

Following decades of conflict and violence, millions of small farmers in Colombia remain economically excluded and trapped in poverty. Many city dwellers and policy-makers are ignorant of their problems, or are not interested. However, an initiative in the capital Bogotá has strengthened links between rural and urban communities and has demonstrated that small producers can become political and economic actors – and that the small producer economy is not only viable, but can be an important part of the solution to urban food insecurity. To achieve this, small producers, NGOs, and municipal authorities have worked together, building a successful advocacy strategy based on a practical working model of access to markets.

Introduction

For the past 40 years, Colombia has suffered from violent conflict involving illegal armed groups who, from the 1980s onwards, have colluded with the country's drug cartels. This has led to gross violations of human rights, fuelled to a large extent by social inequality, poverty, and a lack of economic opportunities in rural areas. Peace talks between the government and the main left-wing rebel group, FARC, collapsed in 2002. The recent process designed to demobilize the paramilitary groups has led to a culture of impunity, the concentration of land ownership,[1] and increasing control of political parties and institutions by members of these groups.[2] The appropriation of land and natural resources is one of the main drivers of the conflict in Colombia. There is a need to distribute at least seven million hectares of land in order for rural households to generate enough income to meet their basic needs.[3]

Together with the political violence, drug-related crime has made Colombia one of the most violent countries in the world.[4] Armed conflict has displaced more than three million people,[5] and thousands have died. Throughout the country – and especially in rural areas – the rule of law is weak, corruption is rife, and human rights violations go unchallenged.

Despite the years of violence, Colombia has maintained long-term economic growth and is one of the largest economies in Latin America. At the same time, however, inequality has risen and it is now the third most unequal country in Latin America and the ninth most unequal in the world.[6] Rural areas in particular have lagged behind. While an estimated 39.8 per cent of Colombia's urban population live in poverty, poor people in rural areas represent 65.2 per cent of the total population.[7] Rural women are particularly disadvantaged,[8] as are as indigenous communities.

To increase their economic alternatives and to boost their incomes, some small-scale farmers (*campesinos*) have turned to coca cultivation. In 2007, the number of households involved in growing the crop increased by 19 per cent to 90,000.[9] In its attempts to control the drug trade, the government has focused its efforts on pursuing the weaker part of the supply chain – the coca growers – while not putting enough effort into capturing cartel members. It is spraying illegal crops, but this has been costly and ineffective, and has damaged legitimate crops and the livelihoods of hundreds of innocent families.[10]

'Because of the fumigations we have lost our crops, especially the plantains which are our main source of food and the crops we sell – papaya, watermelon, and cantaloupes – from which we earned a living. Now three of my seven children have gone to the mountains, to the coca crops,' said one female small farmer of Afro-Caribbean descent, who in order to protect her own safety wished to remain anonymous.

Box 3.1 The lure of the coca trade

'What happened with maize here was that imported maize started to come in at very low prices and we small producers could not lower our prices so much, so this motivated many producers to start to grow coca,' explains one small farmer from Cauca, who wishes to remain anonymous for security reasons.

Another says: 'Since we did not own land, my children had to work on the nearby farms where they earned around $7 a day. If they work harvesting coca crops they can earn $15 a day. But it scares me: if they start working with coca they may join the guerrillas. I am participating in this programme because I am trying to make a better life for my children.'

As well as lacking public investment,[11] rural populations lack a strong voice to speak for them. Although there are some strong indigenous movements, there are few effective civil society organizations (CSOs) or mechanisms capable of holding the state to account. In addition, in the 1990s the economy was opened up to foreign competition without first putting in place measures to protect small farmers, and today huge numbers of people in rural areas are struggling to make ends meet.

In some ways, there is a sense of Colombia being a country of two realities: in more affluent areas of the cities it is possible to live completely unaffected by the brutal conflict, while in rural areas violence, intense poverty, and a widespread sense of exclusion and disempowerment are daily challenges for millions of people.

Making the case for change

Small-scale agricultural producers supply 35 per cent of the food consumed in Colombia,[12] a proportion that rises to 67 per cent in the capital Bogotá.[13] However, the important role that small-scale producers play in the economy is not reflected in their incomes: millions of people reliant on small-scale agriculture live in poverty and more than five million suffer from malnutrition, in one of the most biodiverse and fertile countries in the world.[14]

This is due in part to the weak position of small farmers in marketing their produce. Constraining factors include variable and limited access to land and markets, poor information and quality control, and lack of experience and power in negotiating market transactions. Much of the produce sold in urban markets is handled by commercial intermediaries, who buy from individual producers at low prices and then make a profit without adding any value to the products.

In 2005, Oxfam GB embarked on a programme aimed at helping to overcome the barriers that prevent small farmers from getting fair returns from markets and at creating fairer commercial links between poor rural and urban populations. Its main implementing partner was the Instituto Latinoamericano de Servicios Legales Alternativos (ILSA), which works with peasant organizations and rural communities to protect the rights of poor rural people.[15] The other key partner was the Comité de Interlocución Campesino y Comunal (CICC), a committee that helped to co-ordinate and lead the project.[16] The idea was to support small-scale rural producers, especially women, by strengthening their role in the rural–urban food supply chain. At the same time, the intervention would make good-quality food available at affordable prices for low-income families in urban areas – a win–win situation for both.

Oxfam and its partners wanted to document practical examples that would persuade government that the *campesino* (peasant) economy could become a driver of development, providing a route to economic growth and poverty reduction. By demonstrating that small-scale producers were worth investing in, they also hoped to gain legitimacy to lobby the US Congress on the potentially negative impacts of the US–Colombia Free Trade Agreement, which was being negotiated at the time. Social organization and advocacy to influence public policy were key components of the programme. In 2003 Oxfam supported ILSA's organization of the Agrarian Congress, which brought together more than 5,000 leaders from all over the country and led to the design of the Agrarian Mandate – the first commonly agreed agenda for the rural sector.

Box 3.2 Opening up dialogue, building trust

At the outset of the project, Oxfam and ILSA conducted research to obtain clear data on the potential of the peasant economy.[17] They consulted with academics, political leaders who had worked with peasant movements, experts on rural women and their role in social movements, rural leaders, peasant organizations, INGOs, NGOs involved in agro-ecology, and peasant farmers, both men and women, with whom Oxfam was already working. As well as consulting with potential allies, the programme partners wanted to hear the opinions of those who did not necessarily share their views – for example, government officials who thought that the peasant economy was no longer viable and had been 'sent to the scrap yard'.

Consulting with opponents proved to be very useful, both in terms of gathering information and in gauging the levels of prejudice and mistrust harboured by public institutions and officials towards peasant farmers. Officials frequently expressed opinions such as, 'The peasants are lazy, they support the armed groups, they take advantage of the drug mafias, and take their cut growing coca.'

This made it abundantly clear that the problem was not simply one of information, but also one of trust and goodwill. Oxfam and ILSA realized that their work had to involve building trust between public officials and peasant farmers. As it turned out, this proved to be the single most important factor in the success of the project.

Seizing the opportunity

Colombia's national government showed little interest in Oxfam's proposal. In 2004, Oxfam lobbied the Ministry of Agriculture, but officials claimed that small-scale producers were already covered by its policies. Official data showed, however, that the Ministry considered 'small' producers to be those who owned up to 50 hectares of land. Consequently, Oxfam decided to adopt a strategy that would 'bypass' the national government, instead supporting partners to lobby the food security programmes of the local authorities running Colombia's main cities, such as Bogotá and Cali.

The municipality of Bogotá ranks second in the country in terms of executive power after the national government itself and is Colombia's biggest food market, with seven million consumers. Here Oxfam and its partners were able to capitalize on an important opportunity: in 2004, the Office of the Mayor of Bogotá had begun to speak openly for the first time about poverty levels in the city and the need to make food accessible to all its inhabitants. It had begun to draw up a 'Food Supply Master Plan', but this failed to recognize the potential of small producers.

Neither small producers nor the wider public knew much about the city government's plan, but Jesus Anibal Suarez, the head of ILSA's Agrarian Programme and a leader well respected by peasant organizations, managed to disseminate the main ideas contained in it. The plan was based on obtaining food supplies for the minimum possible price, but it did not consider the needs of producers or the potential for reducing rural poverty. If the plan were to favour large food marketing companies rather than small farmers, it would risk destroying the livelihoods of thousands of small producers reliant on selling their produce in Bogotá. Instead of a minimum price, ILSA, rural organizations, and Oxfam proposed that the plan should be based on a 'fair price' principle.

In formulating its strategy, ILSA drew on past experience of consensus-building and called together the organizations that had participated in the Agrarian Congress. Based on this collective agenda (the Agrarian Mandate), the organizations founded in 2003 the Comite de Interlocución Campesino y Comunal (CICC), a body that united 12 peasant and community organizations with some 1,000 members between them. CICC's aim was to influence Bogotá's food security policy. In forging a broad alliance capable of making change happen, vital lessons were learned about the importance of partners having clear agendas and the need for credible leaders within local communities.

It is rare in Colombia that public institutions and social organizations engage in constructive dialogue, due to mutual distrust, so it was very difficult to convince the Mayor of Bogotá to sit down with the *campesinos* to discuss their proposals for the design of the plan. Nevertheless, ILSA and the CICC persevered, and ILSA was able to bring together more than 100 people at a series of meetings to agree on proposals to lobby the Mayor. As a result, the producers formulated a number of clear demands, designed to fit with the Mayor's own agenda. Their main message was: 'We want to be part of the solution to hunger in Bogotá. Our products can solve the city's food problem.' This clear and positive message was very important in influencing the attitudes of city officials and in encouraging other individuals and organizations to join the campaign.

However, at this point the Mayor's Office was still sceptical of the peasants' ability to supply food in an efficient way. Oxfam, ILSA, and the CICC therefore set out to make the producers' voices heard and to demonstrate their organizational capacity in a creative way. They organized farmers' markets in strategic locations around the city – such as Plaza Bolívar, the city's main square – to coincide with key moments in discussions about the food supply plan.

Around 200 producers participated in each market and drew a great deal of interest from urban consumers, both in the excellent quality of their products and the attractiveness of their prices. Women played a fundamental part in the process, participating in the markets and in campaigning activities aimed at building public support. They showed great determination in overcoming obstacles to market access and boldness in adopting new strategies. One key city government official in charge of the plan who visited a market confessed to being surprised by the level of organization and capacity on display. The campaign demonstrated to officials that the issue was of interest to a broad section of the population.

Forging new alliances

The Mayor's Office was beginning to recognize the potential of small producers, but there was still no proper forum in which to debate the plan. ILSA therefore simultaneously targeted three different, but equally important, actors in an attempt to open up dialogue and influence the design of the Food Supply Plan and the Food Security Policy for Bogotá:[18] the local authorities in the municipalities where producers were located; the private sector company Corabastos, Colombia's biggest food distributor; and Bogotá's Municipal Council, which had the political power to push the Mayor to open up a dialogue with the producers' organizations in formulating the food supply plan.

ILSA found that the municipal councils in the producing areas – who as public bodies could engage in dialogue with officials in the city – knew little about the food supply plan. It persuaded 50 councils, of all political complexions, to sign a joint communiqué calling on the Bogotá City Government to consult with producers about its design. ILSA also held discussions with Corabastos, which handled most of the food sold in shops all over Colombia. This was a bold move, as the firm has the greatest power in the country to set food prices and could be seen as a key actor in preventing small producers from achieving greater power in markets. However, in a groundbreaking tactical alliance, ILSA encouraged the company to use its influence with local politicians, pointing out that it risked losing out itself if the Mayor's Office excluded it from the food supply plan. This was the first time that small producer organizations had formed a tactical alliance with this company.

In April 2005, ILSA arranged for a leader of the *campesinos* to directly address the Municipal Council, the first time that this had happened. Simultaneously, a farmers' market was organized by ILSA, Oxfam, and CICC in the public square outside the Council building, to symbolize the importance of small producers to the city's food security. In his address, the producers' representative invited councillors to visit the market at the end of the session. Also that day, Colombia's highest-circulation newspaper, *El Tiempo*, published a letter signed by the 50 municipal councils stressing the importance of involving small producers in the food supply plan. The result was that, the same day, the Municipal Council committed to push the Mayor to open up the city's food security plan to public discussion. Subsequently, the Council organized a series of public discussions that initiated dialogue between the Mayor and producers in the surrounding municipalities.

At ILSA's suggestion, discussions were held in the producing areas, via decentralized sessions of the Council. This allowed officials to see for themselves how producers lived, and enabled more producers to participate directly in the sessions. After a series of regional consultations, the Mayor of Bogotá finally agreed to revise the city's plan from one based on the lowest possible price to one based on fair prices. The Mayor also agreed to invest in enabling small producers to supply the city in an efficient way. An implementing committee was set up to manage the supply plan and to make decisions on how and when to invest funds; it was agreed that this committee should include two representatives of the producer organizations, giving *campesinos* a direct say in the way that the plan was implemented.

The Bogotá Food Policy was finally approved by the city, covering the period up until 2015. It expressed clearly the need to support small producers to access the city's food market as a means of improving their living conditions, as well as facilitating access to food for poor urban consumers. This new approach has the potential to benefit both the seven million inhabitants of Bogotá and the thousands of producers living in the surrounding areas.

Consolidating and scaling up the initiative

The work was still not finished, however: despite these real achievements, Oxfam, ILSA, the CICC, and the producers wanted to see concrete investments from the Mayor's public budget to turn the promises into real political action. Furthermore, with mayoral elections due across the country in October 2007, there was a risk that the gains and promises might be lost with a change of officials. It was important to ensure that future budgets guaranteed resources and that the supply plan could be demonstrated to be workable.

To help achieve this, the partners pursued a number of strategies simultaneously. First, ILSA and Oxfam supported the CICC organizations to elect their two representatives to the implementing committee of the Food Supply Plan. Second, in partnership with the Municipal Council they set up a pilot marketing project called 'Mercados Campesinos' (farmers' markets). Oxfam played an important role in mediating between officials and the producers' organizations, as at that time there was still much mistrust between them. At last, however, the pilot project was implemented, managed by ILSA. It initially involved more than 1,000 producers organized into local marketing groups, which helped to build capacity and generate economies of scale.

Third, as the elections approached, the CICC targeted candidates who expressed support for their campaign, even convincing some to sign legally binding commitments. Persuaded by the success of the Mercados Campesinos pilot project, most candidates with the best chances of winning a seat included the issue in their manifestos, which helped to create a solid base of support once the new municipal administrations were formed.

Since then, more than 30 municipalities in the area around Bogotá, seeing the success of the markets in the city, have decided to organize their own local markets. Bogotá's Mayor has signed contracts with five other regional governments to increase rural investment for food production. To ensure that these gains are sustainable, ILSA, with support from Oxfam, is lobbying another

40 municipal councils to ensure that they approve public policies supporting small producers, together with technical support and assets for accessing markets.

Following this success, Oxfam aims to scale up the model of smallholder involvement by replicating the initiative in Medellín and Cali, the two other mayoral cities in Colombia. Oxfam has shared its experience with officials in Medellín, who are currently designing their own food supply policy, as well as with the Mayor of Cali. Indigenous organizations in the south-west of the country have initiated a lobbying process similar to the one employed in Bogotá, pursuing dialogue with officials and organizing markets in the city.

At national level, it has proved very difficult for the programme to influence changes in either policy or investment. However, one avenue that offers potential to influence national-level politicians has been opened up through engagement with US Congressmen who have an interest in international trade agreements (see Box 3.3).

Box 3.3 Informing advocacy at the international level

The Colombian Government has attempted to negotiate free trade agreements (FTAs) with partners such as the USA and the EU, but its efforts have failed due to the country's continuing human rights violations and unstable security situation. An FTA negotiated with the USA during the Bush administration has been held up by the US Congress, until the Colombian government can guarantee that the rights (and lives) of trade unionists will be respected.

It is essential to protect the interests of small-scale producers under any such agreement, and Oxfam has lobbied the US Congress to consider the impact that such a deal might have. In a country where 40 per cent of basic foodstuffs are produced by small farmers it would be wrong, argues Oxfam, to approve an agreement that would push them out of the market. Such a move would only serve to fuel armed conflict and the cultivation of illicit crops.

Oxfam's programme in Colombia, through initiatives such as the Mercados Campesinos project, shows that, with the right policies, it is possible to improve producers' incomes and to turn a peasant economy into a driver of development. Oxfam has proposed to the US Congress that the treaty should not be approved as it stands, but instead it should include measures to protect Colombia's small-scale producers and food security in general.

Oxfam and its partners in the USA and Colombia have already shared the results of their successful initiative with Congressmen and have invited them to visit Colombia to see the reality and to hear small producers' proposals for themselves. It is hoped that a practical example of smallholder capacity, with clear benefits, will convince decision-makers of the economic and social importance of small-scale agriculture.

The programme's impact

- As a result of the advocacy undertaken by Oxfam and its partners, a wide social movement representing small-scale farmers is developing, leading to changes in policy, market systems, and consumer behaviour.
- As of December 2009, the Mercados Campesinos project had benefited more than 2,000 producers with less than 5 hectares of land each. Their total sales over three years have been worth more than $3 m. These sales have represented a real benefit for small producers: monitoring by Oxfam and ILSA has shown that the average net increase in prices for farmers is 64 per cent.
- Fifty municipal groups have now been constituted and these groups are growing; the only requirement for a prospective member is to be a small producer willing to sell directly to the city. These local groups collect together farmers' produce and jointly cover the costs of transporting it to the city.
- Urban consumers have also benefited, with prices averaging 34 per cent lower than in large chain supermarkets. This 'win–win' outcome demonstrates the efficiency of producers organizing themselves in order to access urban markets.
- The strategy developed by Oxfam, ILSA, and the CICC played a key role in formulating the Bogotá Food Supply Master Plan and the Food Security Policy, which have been approved until 2015 and are committed to a 'fair price' principle.
- As a result of the advocacy work, the programme has also leveraged an additional investment of $720,000 from the Mayor of Bogotá, with commitments to further develop market infrastructure to boost rural–urban trade.
- The process of working together with the Mayor of Bogotá and building ownership of the action within government has had significant implications for sustainability and for the process of self-replication. The Mayor himself is taking forward the process with mayors in five other regions.
- Women interviewed by external evaluators said that the programme had had a great impact on their daily lives. Most of them had played an active role in advocacy activities and now feel more confident about marketing and selling their produce.

Box 3.4 Real economic alternatives for small producers

'Before, it was difficult just to buy enough food. But now, I have enough money to pay for my children's education,' says one female small-scale producer from Boyacá, a rural area north-east of Bogotá.

'I realize that I can earn a living, that I am capable, and now I am thinking about bigger things. I'm going to begin raising rabbits and chickens as a serious business. Now we know we can do it.'

Challenges and opportunities

Despite the programme's success, challenges remain. There is still an urgent need for other local governments to invest in small-scale producers. A real national policy is needed to guarantee small farmers' access to land and other economic activities, as well as to improve security, so that community members are able to defend their rights without being threatened or even killed.

Social organizations need further strengthening to be able to build on their work in defending the rights of rural people and giving them a voice. There is also a need for effective public forums where the opinions of such organizations are taken seriously and are considered in the design of public policies. In addition, wider campaigns to sensitize public opinion on the reality of life in rural areas have the potential to improve the engagement of consumers (and voters) with the problems that face small-scale producers, and to help in the search for solutions.

Conclusion

One of the key drivers of Oxfam's programme in Colombia was to help create legal economic alternatives for small producers, by strengthening civil society and by tackling poverty and the lack of economic opportunities that lead to inequality and exclusion. Real opportunities for sustainable livelihoods reduce the incentive for people in rural areas to join armed groups or to become involved in illegal coca supply chains.

The programme has provided an example of how to carry out evidence-based advocacy, while achieving real changes in people's lives. It has succeeded in developing important links between partners and producers, has achieved a wider impact through influencing policy processes and political leaders, and has demonstrated the capacity of small-scale farmers – especially women – to gain direct access to urban markets and to participate in better commercial linkages. The outcome has been fairer prices for both producers and consumers.

The process has also provided important lessons for Oxfam and its partners about how change happens. These include the importance of focusing interventions and designing clear but flexible advocacy strategies, aimed simultaneously at authorities at different levels. It has emphasized the importance of engaging in dialogue not only with allies but also with adversaries, and of bypassing national government where this is acting as a 'blocker'. It also shows the importance of working 'where the energy is' – i.e. supporting the agendas of existing social organizations that are able to identify opportunities and capitalize on them when they arise. Strong and accountable social leaders, together with well-organized communities, can help to build connections with decision-makers, resulting in positive changes in public policies.

In summary, the process demonstrates the value of using a well-constructed case to convince others that change is possible.

Notes

1 Just 0.5 per cent of landowners in Colombia own 60 per cent of the land, and statistics show a worrying trend of accelerating land concentration over the past decades. The proportion of properties of more than 200 hectares increased from 47.1 per cent of the total land stock in 1984 to 68.3 per cent in 2000, and to 76.1 per cent in 2005. At the same time, the proportion of properties of less than 3 hectares decreased from 2.9 per cent of the total land stock in 1984 to 1.9 per cent in 2005. Statistics for 1984 and 2000 from Kalmanovitz, S. and López Enciso, E. (2005) 'Tierra, conflicto y debilidad del Estado en Colombia' [Land, conflict and state weakness in Colombia], issue 44, *Observatorio de la Economía Latinoamericana*, based on calculations by Instituto Geográfico Agustín Codazzi. *Source*: CID-UN (2006) 'Bienestar and Macronomía 2002–06, Crecimiento insuficiente, inequitativo e insostenible', Bogotá.

2 Sixty per cent of the Senate consists of political parties close to the government (known as *partidos Uribistas*). Since June 2009, when the process of demobilizing the paramilitaries began, 67 per cent of these senators have been judicially investigated for links with paramilitary groups or drug traffickers. *Source*: Fiscalía General de la Nación y la Corte Suprema de Justicia (General Prosecutor's Office of the Nation and the Supreme Court).

3 CEPAL, see http://www.eclac.org [accessed 16 March 2011].

4 BBC News Country Profile [Online]   http://news.bbc.co.uk/1/hi/world/americas/4528631.stm [accessed 16 March 2011].

5 Ten per cent of the country's population have been displaced at least once. See http://www.accionsocial.gov.co [accessed 16 March 2011].

6 The Gini index increased from 0.544 in 1996 to 0.563 in 2003. 'Encuesta de Hogares' (Government Household Survey) cited in Conpes Social (2005) 'Colombian goals and strategies for achieving the Millenium Development Goals – 2015', Consejo Nacional de Política Económica y Social, Bogotá.

7 Statistics for 2008 calculated by the National Department of Statistics (DANE) and National Department of Planning (DNP) (2009).

8 In 2003, 33 per cent of households headed by women were classified as poor, compared with 25 per cent of households headed by men. *Source*: Misión para la Reducción de la Pobreza y la Desigualdad (MERPD) (2003) based on Encuesta de Calidad de Vida, Departamento Nacional de Planeación, DNP, Bogotá.

9 United Nations Office on Drugs and Crime (UNODC) 'Colombia: coca cultivation survey', June 2006. Cited in ABColombia Group (2009) 'Fit for purpose: How to make UK policy on Colombia more effective', ABColombia Group, London.

10 Despite extensive spraying of coca fields, between 2006 and 2007 coca cultivation increased by 27 per cent (UNODC, op. cit.)

11 For example, in a recent scandal, the media reported that under the current government the Ministry of Agriculture has given away most of the public subsidies earmarked for small producers to relatives of politicians, famous actors, and other members of the elite. *Source*: www.cambio.com.co, 25 January 2010 [accessed 15 March 2011].

12 Forero, J. (2003) 'Peasant Economy and Food System in Colombia: Contributions for the discussion about food security', Bogotá : Universidad Javeriana.
13 Plan Maestro de Alimentos de Bogotá (Food Master Plan for Bogotá) (2006) Alcaldía Mayor de Bogotá (Bogotá Mayor's Office), Secretaria Distrital de Desarrollo Económico, Bogotá.
14 FAO (2005) 'The State of Insecurity in the World'. Statistics for 2000–02, FAO, Rome.
15 Latin American Institute for Alternative Legal Services. See http://ilsa. org.co:81/node/141 (English); http://ilsa.org.co:81/node/155 (Spanish) [accessed 15 March 2011].
16 Peasant and Rural Community Dialogue Committee.
17 Yepes, D. et al. (2005) 'Consumo de Alimentos en Bogotá. Déficit y Canasta Básica Recomendada (Food Consumption in Bogotá. Deficit and Recommended Food Basket'), ILSA.
18 'Food Supply Plan' refers to the plan for modifying the supply chain and refers to the way in which food is delivered to consumers (accessibility). The plan is part of the city's Food Security Policy, which is designed to ensure accessibility of food for consumers, as well as availability and nutritional value.

About the author

Aida Pesquera is Country Director, Guatemala, Oxfam GB.

CHAPTER 4

Engaging smallholders in value chains – creating new opportunities for beekeepers in Ethiopia

Shekhar Anand and Gizachew Sisay

An initiative in the Amhara region of Ethiopia has capitalized on the potential of local honey production to build a promising alliance between smallholder farmers and a private sector export company. A coalition of facilitating partners has developed the value chain for honey and other bee-derived products by providing producers with technology inputs, training, and extension services, helping them to organize their production, and creating an enabling policy environment. Farmers who previously produced small quantities of low-quality honey have quadrupled their output and are now producing certified organic honey for export to international markets, which has significantly increased their incomes.

Introduction

Ethiopia is one of the poorest countries in the world, in 2010 ranking 157th out of 169 countries in the Human Development Index.[1] An estimated 77.5 per cent of the population live on less than $2 per day and the adult literacy rate is just 35.9 per cent.[2] Beekeeping is an important economic activity, with the sector contributing around $1.6 m annually to the national economy.[3] The production of honey and beeswax provides a secondary source of income for smallholder farmers, who traditionally also grow cereals, pulses, oil seeds, and chillies. The country has more than ten million beehives, the largest number in Africa, and around two million people are involved in the value chain.[4] Ethiopia is Africa's largest producer of both honey and beeswax, and the fourth-largest producer of beeswax in the world.[5]

Ethiopia has the potential to produce 500,000 tonnes of honey and 50,000 tonnes of beeswax per annum, but currently production is limited to 43,000 tonnes of honey and 3,000 tonnes of beeswax.[6] Moreover, the quality of Ethiopian honey is generally poor, as 95 per cent of beekeepers follow traditional beekeeping practices with no improved techniques or technology.[7] Most honey (over 97 per cent of production) is sold via formal and informal domestic spot markets, and 85 per cent of this is purchased by brewers of *tej*, a honey wine (mead).[8] Income from the sector is minimal, primarily due to low productivity and poor quality, but also because of limited market access, which forces producers to sell locally at low prices. Smallholders produce on average 5 kg of honey per year from each hive, and must travel long distances to markets or sell at low prices to middlemen or local traders.[9]

Globally, there is large and growing demand for honey and other bee products.[10] From 2001–05 the average annual growth rate of honey production globally was 2.3 per cent,[11] although since then supplies have decreased, mainly due to the growing incidence of colony collapse disorder (CCD) in Europe, the USA, and South America.

There is a large unmet demand for organic honey in European countries and, according to the International Trade Centre (UNCTAD/WTO), East Africa has good potential for organic beekeeping. In the past few years, increasing demand has provided Ethiopia with opportunities to export small amounts of smallholder-produced honey to neighbouring countries such as Yemen, Djibouti, and Israel. It has also begun tentative moves to export honey to the European market, and is on a list of approved suppliers to the EU.

Starting out: the pilot project

In 2003, Oxfam initiated a three-year pilot project to promote the trade of honey and bee products in Amhara National Regional State. The evaluation of this programme coincided with the establishment of the Ethiopia Agricultural Scale Up programme, which ambitiously aimed to increase the incomes of one million smallholder farmers through empowerment and better access to market opportunities. As part of its approach, the programme identified specific commodities in which the development of value chains would offer opportunities to leverage private sector investment for poverty reduction and generate evidence to support advocacy activities related to regional and national rural policy.

The honey value chain in Amhara was selected as one of the focus commodities with potential for scale-up because it offered good potential for reducing poverty amongst smallholder farmers, particularly women and landless people. Beekeeping is one of the most sustainable livelihood options for landless people, as landowners pay beekeepers to set up hives on their land to enhance crop pollination. Generally, women do not own land in Ethiopia, and so over 50 per cent of the beekeepers targeted by the programme were women.

The regional government had already identified 31 districts with potential for commercial beekeeping and had developed a plan both to support beekeepers in the use of modern technology and to strengthen the capacity of farmers' unions. To take advantage of this opportunity, Oxfam set out to modernize traditional beekeeping practices and transform small-scale, low-value production into a model of commercialized beekeeping. As people can keep bees in their spare time at home, this plan had good potential to involve women, as well as offering farmers the opportunity to diversify income sources.

Furthermore, honey produced by smallholders is organic and environmentally friendly, and receives priority support from the Ethiopian government as a high-value commodity.[12] Oxfam's programme capitalized on this by developing co-ordination groups, involving government offices at the local, regional, and national levels, as a forum for honey value chain stakeholders to discuss bee products. Additionally, the experiences of beekeepers were shared at an annual national learning event organized by Oxfam in close collaboration with the Federal Ministry of Agriculture and Rural Development.

Box 4.1 New markets, new income

One key objective of the pilot project was to identify more profitable markets for the farmers' growing honey supply. The project staff dedicated significant time to identifying potential buyers before engaging commercial agents Century Trading Ltd and Beza Ltd, who promoted retail-packed honey to over 200 supermarkets and grocery outlets in Addis Ababa.

An evaluation report conducted in 2006 showed that, prior to the project, the price received by producers from local traders was Ethiopian birr (ETB) 5–6 ($0.30–0.40) per kg for crushed honey. Traders also often cheated producers on weights. Through the co-operatives, the producers now receive ETB 32 ($2.40) per kg for Grade 1 pure honey and ETB 28 ($2.15) per kg for Grade 2. Co-op members also receive dividends on the sale of processed bee products to the agents, Beza and Century. These dividends range between ETB 35 ($2.60) and ETB 674 ($50) per season, based on the number of shares the producer owns. Most producers are re-investing their dividends to expand their operations and also the processing centres.

'Before the co-operative, we used to sell our honey at a low price to middlemen in the market. I was perhaps getting ETB 8–10 ($0.60–0.75) per kg, but now through the co-operative we are getting ETB 32 ($2.40) per kg. This is providing 40–50 per cent of my yearly income and covers nearly half of my family's needs, including school and medical fees. With this I am able to send eight of my children to school', says Ato Workneh Addis, a beekeeper from Bahirdar Zuria.

Sources: Terminal Evaluation of the Bee Products Trade Promotion Programme (August 2006) and M. Shepherd (2007) 'Honey Co-operatives Case Study', Oxfam GB (November 2007).

The pilot project aimed to demonstrate the potential of commercial beekeeping for smallholders and was implemented by partner SOS Sahel,[13] with funding from Oxfam. The project worked with this partner to support the development of six primary co-operatives, involving 2,100 farmer households, and created the Zembaba Beekeeper Co-operatives' Union to co-ordinate their activities. Through these organizations, it provided training in production techniques and the use of improved technology, notably the Kenyan top-bar hive,[14] which enabled women to become more involved in honey production (previously honey production involved tree climbing, which was not seen as socially acceptable for women). Local government offices for agriculture and the promotion of co-operatives were fully involved in extension services and in building the capacity of the co-ops. Evaluation of the pilot scheme showed that, on average, productivity improved from 5 kg of honey per hive per year to 20 kg per hive per year.

Training was also provided in processing and, together with the government, the project constructed honey collection and processing centres in eight villages. Previously producers had to travel long distances to find buyers for their honey, which reduced their negotiating power and further excluded women from the market. The project also supported co-operatives to apply for organic certification and provided training in marketing, management, and business skills. A credit facility was established and managed by the Meket Micro-Finance Institute (MMFI), which enabled the co-operatives to purchase honey from member beekeepers through the collection and processing centres.

Oxfam GB's target was to scale up its promotion of good practice in the beekeeping sector to reach 40,000 beekeepers in 20 districts. However, this would require further investment to convert traditional beekeeping techniques to modern ones, develop market linkages, diversify products and opportunities, and improve both institutional capacity and the provision of financial services, with a particular emphasis on women beekeepers.

Scaling up: engaging with the private sector

Oxfam's intention was to build a scalable model that would link small-holders with formal markets. In the initial phase of the programme it was actively involved, providing the co-operatives with services such as training on marketing, contract management, and negotiation skills, together with technical support and periodic monitoring of quality standards and production. It also undertook market assessments and feasibility studies on bee products, developed new market strategies to promote the producers' own-brand 'Amar' honey and beeswax, and facilitated market opportunities for bee products.

However, in the scale-up phase Oxfam's plan was to act solely as a facilitator, and to identify private sector and government actors who could invest in services to develop the model. It was felt that private sector engagement would provide access to larger and more stable markets both within Ethiopia and for export, and would also leverage greater investment to develop the co-operatives' processing and marketing activities. A stakeholder engagement process identified a private company, Ambrosia, as a promising partner with whom a joint programme could be developed. The company displayed a commitment to social responsibility, with a stated mission to create jobs and improve incomes for farmers. It had also already carried out a nationwide feasibility study and quality assessments of Ethiopian honey.

Owned by three Ethiopian entrepreneurs and a French-based company, STECA, Ambrosia was established in 2000, with initial capital of Ethiopian birr (ETB) 54 m ($5.75 m). The company has signed an agreement running to 2011 to supply CSV International, a honey buyer in France, and on the strength of this contract has established the largest honey processing plant in Ethiopia at Debrebirhan, using working capital of ETB 21.5 m ($1.6 m) from the Development Bank of Ethiopia. The plant, which is being fitted out with imported machinery, is the only one in the country capable of exporting processed honey and has the capacity to process and pack 6,800 tonnes of honey and beeswax every year. Ambrosia's ambition is to be the market leader for bee products in both the Amhara region and Ethiopia as a whole.

In 2008 Ethiopia received Euro Gap accreditation[15] to export organic honey to Europe, and subsequently Ambrosia won an order from CSV International to export honey at a price of €3 (ETB 50) per kilo. However, due to the lack of an organized local supply, Ambrosia was unable to fulfil this export contract, which called for 2,300 tonnes of honey a year.

As a result, the company realized that the best way to ensure consistent supplies of honey was to invest directly to establish formal smallholder production, rather than depend on unreliable open wholesale markets. It estimated that it would require 42,000 beekeepers to ensure adequate supplies to meet its needs. It also realized that it would only be possible to reach out to so many farmers through organized beekeeper co-operatives and unions. In addition, it would require a coalition of training institutions, service providers, and facilitating NGOs to help develop efficient small-holder enterprises.

Oxfam and Ambrosia agreed to jointly develop a long-term business plan to systematically expand the operations established under the pilot beekeeping project. Together they established a training school for beekeepers in Mecha district with a joint investment of £66,000 ($105,000). Until now, the partners have focused on establishing sustainable supply chains, but Ambrosia has recently signed a memorandum of understanding (MoU) with the co-operatives and beekeeping union to purchase and market their produce, and was collecting honey in large quantities to start bulk processing in March 2010. Working at full capacity, with three shifts per day, its plant has the potential to produce 6,800 tonnes of honey and beeswax annually, although initially the target will be 2,200 tonnes. Once production has begun, the company hopes to offer farmers who successfully complete its training courses soft loans to purchase improved beehives and modern beekeeping equipment.

Box 4.2 Investing jointly to train producers

There are an estimated 20,880 bee colonies in Mecha district, 95 per cent of which are kept in traditional hives. Oxfam GB, in collaboration with Ambrosia, has set up a training and demonstration school here with the aim of trans-forming subsistence production into commercial beekeeping. Ambrosia has invested £34,000 and Oxfam £32,000. One hundred beekeepers have been trained in three rounds each for five days and will receive improved beehives on long-term loans that will be paid back as honey is sold. The school also provides technical support and other inputs such as honey extractors, veils, and foundation sheets, on a regular basis. As a result of the lessons from the training school, Ambrosia is now planning to work with 42,000 new beekeepers, linking them with formal markets.

Under the MoU, Ambrosia has undertaken to purchase 3,400 tonnes of honey annually (although this may not have been fulfilled in the first year) via the Zembaba Union. By early 2010, while the first harvest season was still in progress, Zembaba had collected over 100 tonnes. The purchase price for the honey is decided by a regional committee at the start of each month. This committee includes representatives from the union, Ambrosia, the Co-operative Promotion Agency (CPA), the Bureau of Trade and Industry, the Marketing Department of the Bureau of Agriculture and Rural Development, and Oxfam GB. When setting the price, the committee takes into account honey prices in different markets in the region and other factors such as production costs. Both parties have the right to reject the price set by the committee if they think it is unfair. Ambrosia collects the honey (using its own transport) when the volume gathered has reached at least 5,000 kg. Once a co-op has proved itself to be a reliable supplier, producers are paid up to 30 per cent in advance, with the balance paid within a week of the honey being received from the union.

The role of facilitator: working with partners

Oxfam took a lead role in analysing and planning the development of the value chain, acting primarily as a facilitator. It also played a key role in ensuring that information was shared on an ongoing basis and that different actors worked together to solve problems. It has helped to develop sustainable relationships between partners such as local agricultural extension offices, the Co-operative Promotion Bureau, the regional Bureau of Agriculture and Rural Development, the Ethiopian Quality and National Standards Authority, and research agencies. A regional taskforce has been established to act as a forum where producer organizations, co-ops, and the union can discuss issues with stakeholders from government and private companies. The taskforce also advises on strategy and lobbies to influence change at regional level (for example, it has persuaded the Bureau of Agriculture to take action on pest control).

As well as buying honey, Ambrosia is investing in training and marketing services (collection, transportation, packaging, finance, and market information), while Oxfam and other partners are responsible for targeting new farmers for beekeeping training and co-operative membership; strengthening the capacity of producer organizations; research and value chain analysis; and identifying opportunities to add value to products (branding, certification, and standards). Oxfam has also established partnerships with a wide range of stakeholders including government departments, local NGOs, research agencies, and producer organizations. Actors who have played crucial roles in the value chain include the following:

- The Zembaba Beekeepers Co-operatives' Union has organized beekeepers under one umbrella, working with individuals to boost their productivity while building the capacity of primary co-operatives and taking responsibility for quality control, processing, and marketing.
- The regional Co-operative Promotion Bureau – a government body – provides co-operatives with technical and managerial support. Part of the regional taskforce, its role is to organize farmers into co-operatives and to facilitate the flow of market information.
- Another government organization, the Bureau of Agriculture and Rural Development, has provided extension services to beekeepers via local development agents.
- Oxfam GB has worked closely with the government's Agriculture Research Institute to provide information to beekeepers on disease and pest control, especially of the *Varroa destructor* mite, a parasitic pest of honeybees.
- NGO partners such as SOS Sahel and ORDA[16] have helped to build the capacity of farmers' organizations and have provided technical support on production and marketing. In particular, the project has worked with the Quality and Standards Authority of Ethiopia on packaging and organic certification.[17]
- SNV Netherlands (the Netherlands Development Organization)[18] has provided technical support on business planning for the Zembaba Union, and has worked to facilitate organic certification, which is essential to gain access to new markets. As well as being part of the regional taskforce, it helps to co-ordinate (with the Chamber of Commerce) a national-level steering committee to help develop the beekeeping sector nationally. A national regulatory framework would eventually allow Oxfam and other partners to exit the intervention, but more work on this is needed; discussions with the Federal Ministry of Agriculture are ongoing.

Achievements to date

From six co-ops in the pilot phase, the programme now involves 36 registered co-operatives with approximately 10,500 members. Of these organizations, nine are members of the Zembaba Beekeepers Co-operatives' Union. Eight village-level honey collection and processing centres have been constructed under the six pilot co-ops. The Zembaba Union serves as a hub for beekeeping development in Amhara, and has built capacity by improving members' skills and knowledge, which has helped to increase productivity by up to 400 per cent and to boost incomes by 200–400 per cent.[19] Increased incomes have enabled smallholders to invest in education and other services.

Most beekeepers have switched their bee colonies from traditional log hives to transitional Kenyan top-bar types, while some have switched to modern frame hives. Some have also started producing beeswax, which offers additional opportunities to increase income. The Bureau of Agriculture and Rural Development and the Regional Agriculture Research Institute, which makes wax foundation sheets on which bees can build honeycombs, currently purchase the wax. Beeswax could potentially be produced for sale to cosmetics and pharmaceutical companies in future, but this needs further organization.

As well as expanding the number of processing and collection centres, the union has developed a market information system, linking smallholders to markets and improving their awareness of quality, demand, and prices. This improves farmers' negotiating power, allowing them to meet buyers' requirements and identify the most profitable markets for their produce.

Box 4.3 Building capacity through the beekeepers' co-operative

'We have capital of £4,300 [$6,865]. Our total membership has increased to 2,600', says Tenawu Mehiratu, manager of the Zembaba Beekeepers Co-operatives' Union, based in Bahir Dar. 'Our union supports beekeepers on technical training and facilitating markets. Now the farmers know about markets, and how to get better prices. Before, 1 kg of honey sold for 30–40 cents (ETB 5–6); now with improved quality, farmers can sell for up to $1.60–$2.40 (ETB 21–35) per kg.'

Mehiratu recognizes that the union still needs to work with farmers to increase awareness of improved beekeeping techniques, quality, and production. 'We need policies to restrict the use of chemicals and to facilitate retailing. This year we have taken a loan of £3,500 [$5,588] to buy honey from farmers. We have 40 quintals [4,000 kg] of honey in stock', he says.

Through the link with Ambrosia, the producers are able to access potentially larger-scale international markets. However, the co-operatives know that dependence on a single buyer is very risky for farmers. So although Ambrosia is a key player in the honey value chain, the co-operatives continue to market their own 'Amar' brand organic honey through local traders and supermarkets. In addition, some beekeepers still sell their honey at informal cash markets. While prices at these markets tend to be lower and more variable they are still a useful outlet, especially for women looking to meet immediate household needs.

Since the beginning of the pilot project, it has been a goal to increase the participation of women in the beekeeping sector. This has been a challenging task, but growing numbers of female beekeepers have learned how to manage the improved hives and beekeeping tools and equipment, through the introduction of new hives and targeted training. However, women's levels of participation are still relatively low and this is a priority for further action. For example, the union and the co-operatives are changing their rules to allow two members per household to join; previously, it was only one member per household, and this was nearly always a man. New training on technology and marketing will target women, and women's beekeeping groups will be formed. Increasingly, also, female beekeepers are taking leadership roles in existing co-ops.

Box 4.4 New economic opportunities for women

Although there are few specific cultural barriers in Ethiopia to women becoming beekeepers, the sector is traditionally male-dominated. When Oxfam began its intervention in Amhara region, it found that only 1 per cent of beekeepers were female, although 24 per cent of households were female-headed. Building the honey value chain promised significant opportunities for greater economic empowerment of women.

The introduction of new technologies and new types of beehive has benefited both men and women, but in particular it has helped to involve more women. The new hives are easier to handle and can be kept at the homestead, where they are easy to access. They are not kept in trees like the traditional types, and beekeepers' gloves and veils have reduced the risk of being stung when harvesting the honey. This has helped to increase women's confidence.

Elements of the programme have been designed specifically to benefit women. For instance, establishing honey collection centres in villages has given women better access to information and has allowed them to engage in marketing. Previously they had to make long, unsafe journeys to distant markets. Women were often not allowed by their husbands to leave their village to attend training sessions, but this problem has been solved by setting up training centres in villages, and providing training at times when women can attend.

The new value chain has also created employment opportunities: for example, women are making the specialist overalls, gloves, and veils required to handle bees. They are also involved in processing honey at the co-op collection centres.

Over the past two years, women's engagement in beekeeping and their role in markets have increased significantly, accompanied by increases in income of up to 100 per cent. The number of female co-op members has increased by 25 per cent and women now account for 17 per cent of total membership. More women are involved in leadership and governance of the co-operatives, and this is likely to increase as a result of new initiatives taken by the co-ops and the union.

Enabling factors and constraints

A number of positive factors have helped to enable development of the value chain and to scale up honey production in Amhara. These include:

- **Market demand:** Demand for the product and strong motivation to exploit the supply opportunity presented by smallholder agriculture were essential prerequisites in establishing the business case for private sector investment.
- **Suitability of the product:** Local knowledge was important in commercializing smallholder production, allowing appropriate techno-logical improvements to be made to improve the quality and scale of production. With suitable flora and environmental conditions, small-holders were able to meet the requirements of organic production and now have opportunities to access niche markets as international demand continues to increase.
- **Policy environment:** The scale-up was directly linked with the policy environment. Such projects can only be replicated if minis-tries and local authorities support the value chain with policies on the provision of market services, trading, certification, and revenue that allow private companies to enter into service provision and contract farming arrangements. In Ethiopia, beekeeping has been identified by the government as a high-value sector. Commitment from the government, the potential of production, and high market demand all encouraged actors in the chain to come together. The regional taskforce identified additional advocacy issues related to production, quality, and pricing.
- **Flexibility in working methods:** Engagement with the private sector requires the building of confidence between smallholder farmers and private companies. Typically smallholders lack organization and access to communication, and this gives market players a negative picture of

their strengths and credibility. In engaging with Ambrosia, dialogue was made possible by Oxfam's facilitation through value chain forums and by increased communication through information sharing and visits to project sites. In addition, the farmers' union had already been established to represent the primary co-operatives, and with the assistance of Oxfam and its partners was able to build a credible relationship with the company.

- **Ability to link with diverse markets and service providers:** Dependence on a single purchaser creates a high level of risk for smallholders. If that company stops buying or providing services, it can have a catastrophic impact on smallholders' incomes. In this case, it was possible to work with government agricultural extension services, the co-operatives union, and Ambrosia on service provision. The agreement with Ambrosia has helped to improve production and the quality of the honey, as well as guaranteeing a market. The co-operatives are also developing their own brand and exploiting domestic and regional markets as well as European markets through Ambrosia.

The relationship of producers' organizations with Ambrosia is based on value chain principles (not as in a supply chain where the buyer dictates terms to suppliers). Agreements will be effected if both parties see benefits in them. Producers can break off relationships with any private sector actor if they feel that these are too risky. As producer organizations are autonomous, they do not face the risk of being dependent on a single buyer. Oxfam is working to help the beekeepers' union and co-operatives reduce such risks by 1) ensuring that the agreement with Ambrosia is fair, transparent, and based on value chain principles; and 2) building the producer organizations' capacity to identify alternative markets.

Despite the opportunities created, there are still some factors limiting the potential to achieve wider change:

- **Inexperience in export markets:** Farmers' organizations are knowledgeable about immediate local markets, but a lack of experience in accessing international markets is a constraint on their profitability. However, in the scale-up phase of the programme, the union is undertaking a planning exercise to develop a viable business plan over a number of years. Potentially interested buyers from the Middle East have started negotiations to buy table honey, and discussions have taken place with regards to long-term engagement with the Fair Trade Labelling Organization (FLO), with further action planned.
- **Maintaining quality and supply:** The growing number of farmers in the beekeeping sector poses the challenge of maintaining democratic systems and governance of farmers' organizations, which have a direct bearing on quality and supply. The capacity of co-operatives can affect production levels and the quality of produce, for example.

- **Limited exploitation of other bee products:** To date, smallholders have focused on the production and marketing of honey and beeswax. Other bee products for which there is high potential market demand – such as propolis and venom – are not currently being considered due to lack of technology and skills.
- **Financing and policy constraints:** There are currently only limited financial services available from mainstream financial institutions for co-operatives and unions, while trade barriers can prevent small producers from developing their own brands and market promotion strategies.
- **Environmental degradation:** The beekeeping sector is dependent on healthy flora and a healthy environment. Recent years have seen environmental changes in Ethiopia in terms of erratic rainfall patterns and deforestation. If these problems worsen, the beekeeping sector could be affected.

Conclusion

Oxfam's work on the honey value chain has been successful in demonstrating good practice in the commercialization of beekeeping in terms of improving productivity, organizing producers, strengthening farmers' capacity, and linking farmers with formal markets. The intervention has reduced poverty among smallholder farmers, particularly women and landless people. The organization of farmers, the introduction of new technology, the equal participation of women, adding value to produce, awareness and information on markets, and linkages with the private sector are all factors with potential to sustain growth of the value chain. It is vital there continues to be a body that can play a facilitating role between the different actors involved in the honey value chain. Ultimately Oxfam's exit strategy must enable market players to take over this role. It has therefore avoided continued subsidization of market services in order to allow an eventual exit.

This case demonstrates that private sector linkages can clearly benefit smallholder farmers. However, farmers' organizations need to understand market mechanisms and build their capacities accordingly. Maintaining levels of supply and the quality of produce are key priorities if farmers are to benefit from formal markets. Organization through value chain forums can play a crucial role in addressing policy issues, through joint action by farmers and their partners. Finally, the scaling up of honey production must be incorporated into national agricultural policy, so that Ethiopian beekeepers can realize their full production potential.

Notes

1   UNDP (2010) *Human Development Report 2010*, [Online] http://hdr.undp. org/en/statistics/ [accessed 16 March 2011].

2 UNDP *Human Development Indicators*, [Online] http://hdrstats.undp.org/en/countries/profiles/ETH.html [accessed 16 March 2011].

3 Ministry of Agriculture and Rural Development, *Overview in Apiculture in Ethiopia*, Government of Ethiopia, Addis Adaba, December 2008.

4 Ibid.

5 Allafrica.com (2009) 'Ethiopia: land of wax and honey', [Online] http://allafrica.com/stories/200904170706.html [accessed 16 March 2011].

6 Ministry of Agriculture and Rural Development, *Overview in Apiculture in Ethiopia*, Government of Ethiopia, Addis Adaba, December 2008.

7 Oxfam (2008) 'Partner progress report: SOS Sahel', Oxfam International, Bahir Dar. The honey produced in traditional hives is often mixed with wax, pollen, dead bees, and extraneous matter. This means that it cannot be used for processing or for export as table honey, but is only suitable for use in *tej* brewing.

8 Oxfam (2008) 'Partner market report', Oxfam International.

9 Oxfam (2008) 'Partner progress report', Oxfam International.

10 Bee venom is used in medicinal products, royal jelly in medicines and beauty products, and propolis in medicines, varnishes, and chewing gum.

11 Oxfam GB (2009) 'Market and value chain analysis report', Oxfam, Addis Ababa.

12 Amhara Region Agriculture Bureau, Five-Year Strategic Plan (2005–2010) for Apiculture Resource Development, February 2005.

13 See http://www.sahel.org.uk/ethiopia.html [accessed 16 March 2011].

14 Traditional hives are typically made of hollow logs sealed with mud and sticks or reeds; the ends are opened to harvest honey, which disturbs the bees and damages the combs. The hives cannot be inspected without breaking them open. Top-bar hives consist of a simple box (which can be made of local materials by beekeepers themselves) with wooden bars that can be lifted and inspected individually. Modern hives are rectangular boxes with frames inside that can be drained of honey and then returned with the comb intact. These require precision carpentry work to construct.

15 For free trade with European countries. Euro Gap accreditation is granted only after the applicant has fulfilled the strict requirements and standards of the European Union Trade Commission. These include high production standards, quality assurance, and sound environmental practices. Food, Agriculture and Natural Resources Policy Analysis Network (FANRPAN). See http://www.fanrpan.org/ [accessed 16 March 2011].

16 Organization for Rehabilitation and Development in Amhara (ORDA), see http://www.ordainternational.org [accessed 16 March 2011].

17 Quality and Standards Authority of Ethiopia, see http://www.qsae.org [accessed 16 March 2011].

18 See http://www.snvworld.org/en/Pages/default.aspx [accessed 16 March 2011].

19 Oxfam GB (2008) 'Project monitoring report', Oxfam, Addis Ababa.

About the authors

Shekhar Anand is Programme Director, Ethiopia, Oxfam GB.

Gizachew Sisay is Senior Value Chain Advisor, Oxfam GB.

CHAPTER 5

Power to producers – building a network of dairy enterprises owned by local farmers' groups in Haiti

Luc Saint Vil

Contributors: Hugo Sintes and Claire Harvey

Lèt Agogo, a project initiated and led by a Haitian NGO, Veterimed, has created a network of small dairy enterprises that process milk and cheese products for domestic markets. Key to the project's success has been the involvement of local farmers' associations, who supply the dairies with milk; in return, farmers receive technical and material support and benefit from collective inputs and marketing. As a result, their incomes have risen and women in particular have been able to play a more prominent economic role. The next step is to scale up the network nationally, using a franchise model that will enable shared ownership between producers, processors, and the brand holder, and which will transfer management responsibility for individual dairies to the producers themselves.

Introduction

This chapter tells the story of the Lèt Agogo project from its inception in 1999 up until 2009, but was largely written before the devastating earthquake that struck Haiti on 12 January 2010. A brief update on how the project has been affected by the earthquake and plans for taking it forward during the period of recovery come at the end of the chapter.

Even before the earthquake, Haiti was one of the poorest countries in the world. Political instability, civil war, natural disasters, deforestation, and economic liberalization policies have all contributed to severe structural poverty. The country ranks 149th out of 183 countries on the UNDP's Human Development Index, with 72.1 per cent of its population living on less than $2 a day and 54.9 per cent living on less than $1.25 a day.[1]

Despite the fact that over 75 per cent of Haiti's population work in agriculture or depend on the sector for their living,[2] agriculture has undergone a serious decline in the past two decades, with both production levels and the volume of agricultural exports falling.[3] The continuous political turmoil and the neoliberal policies implemented in the late 1980s have contributed to a significant decrease in national production and have exacerbated the food insecurity of poor and marginalized people. In 2007, the World Food Programme (WFP) estimated that 25 per cent of rural households were food insecure.[4] Import taxes have been reduced or even abolished for the majority of agricultural products. As a result, local markets have been flooded by subsidized products from industrialized countries and local producers have struggled to compete.[5] In the dairy sector, all of Haiti's milk processing factories had closed down by the mid-1990s.[6]

Small farmers in Haiti typically farm less than five acres of land, with no irrigation and only basic tools. Serious investment in agriculture is needed to raise the productivity of small farmers in order to reduce food imports and replace them with domestic production. Promoting new areas of local production – such as dairy products – is therefore a key opportunity for boosting income and increasing food security.

The idea behind Lèt Agogo

Annual consumption of milk in Haiti is estimated to be 130,000 tonnes, of which 45,000 tonnes are produced domestically.[7] However, with between 450,000 and 675,000 dairy cows, potential national production is estimated at 145,000 tonnes.[8] Due to a lack of technical ability and adequate infrastructure for processing and marketing, 100,000 tonnes of milk are wasted every year. Some farmers sell milk to merchants, but prices are generally poor and markets unreliable. Many farmers do not even milk their cows because they are not able to sell the milk immediately or preserve it. Meanwhile, milk and dairy products are the second biggest category of food imports, with more than €40 m ($59 m) worth of evaporated milk, UHT milk, yoghurt, and cheese coming into Haiti every year.[9]

Haitian NGO Veterimed is attempting to confront this challenge through its Lèt Agogo dairy project, which is working to provide a secure outlet for small producers' milk. Established in 1999, Lèt Agogo has built a national network of 13 micro-dairies that process milk produced by local farmers' associations into a number of different products – sterilized bottled milk, pasteurized milk, flavoured yoghurt, and artisanal cheese – which are sold for profit through various outlets (street sellers, supermarkets, and the government's school programme) across Haiti. As well as farmers' associations, the project involves the participation of other NGOs and agencies (including Christian Aid, FAO, the European Union, and Veterinarians without Borders) and youth and women's groups, including the African Institute for Development Policy (AFIDEP). It receives financial support from a number of sources, but has been supported from the beginning by Oxfam GB.

Veterimed has followed a shared ownership franchise model in building the Lèt Agogo network. Investors are identified to support the construction of a dairy, by purchasing equipment and recruiting staff. The dairies use basic and easy to maintain equipment that is appropriate to the local context and which requires a relatively small initial outlay. Each dairy employs between five and 15 staff.

Box 5.1 How Lèt Agogo works[10]

Every day farmers and their children travel by foot, bicycle, or donkey to deliver their fresh milk to Lèt Agogo dairies across Haiti. Here, dairy staff test the milk rigorously for dilution and impurities, ensuring that it is of high quality. The dairies handle close to 150 gallons of milk per day. They are well adapted to the lack of electricity in Haiti, transforming milk into cheese, yoghurt, and sterilized bottled milk via simple processes that require only the use of propane gas and running water.

These products have average profit margins of $0.02 per bottle for yoghurt, $0.10 per bottle for sterilized milk, and $1.80 per pound for cheese. The sterilized milk is often flavoured with vanilla, chocolate, or strawberry, and can keep fresh in heat-resistant bottles for up to nine months without any need for cooling.

In some areas, dairies represent a first ever opportunity for producers to sell milk commercially. Richard Joseph, the president of a producer association in Terrier Rouge, explains: 'Before the dairy existed, I had cows at home but I didn't sell much milk at all, because local buyers weren't offering good prices. But since we've had a dairy in our community, I know there's a reliable place I can take it for sale.'

The dairies have supply contracts with government schools, which before the earthquake bought between 1,500 and 5,000 bottles of milk each week. This meant that dairies could offer producers fixed prices: $1.50 per gallon as opposed to $1.00–$1.40 on the street. Unlike street buyers, the dairies could also offer producers a guaranteed regular income. Of producers surveyed, 89 per cent reported that their annual incomes had increased since joining the Lèt Agogo network, while only 26 per cent of non-beneficiary producers reported similar income increases.

The majority of producers have used the new surplus income from milk sales to pay for their children's school fees. One producer in Terrier Rouge was able to pay for tuition for seven of his children and then to send two to university. He said, 'Were it not for milk sales, they wouldn't have been able to attend.'

Many producers collect their earnings after several months and invest in items such as motorbikes and livestock, which they can use to earn further revenue. Some have simply been able to improve their living conditions, repairing their roofs or building new houses.

Most importantly, the model requires that dairy farmers living nearby are recruited to form a producer association, which co-ordinates milk production and delivery to supply the dairy. Veterimed provides capacity-building support and technical training for association members and, depending on funding, may also provide other inputs such as water wells or cows (cows are provided specifically to enable women to get involved). Each dairy has a board of directors that includes farmers' representatives.

A Veterimed-funded marketing and purchasing 'hub' called the Central Purchasing and Commercialization Unit – the 'Central Unit' – co-ordinates marketing and sales for all dairies, finding clients for the products, negotiating contracts, and helping the dairies to restock equipment and packaging supplies – a service for which it takes a small margin to cover its costs. The individual dairies then deliver the products to clients, claim a small margin to cover their own expenses, and pass on a premium to the producers.

In addition to processing the products, Lèt Agogo aims to increase dairy productivity by improving animal health services and feed quality; reduce animal mortality rates; provide organizational support to farmers' associations; and promote the consumption of dairy products at the national level. Its dairies now claim an estimated 0.4 per cent of the national milk market and just over 1.1 per cent of the market for locally produced milk.[11] Lèt Agogo's products are the only dairy products currently on the market that are made from locally produced milk.

Creating the potential for scale

The franchise model has been promoted to enable self-replication of a profitable business across the country. The programme staff identified the following key factors in ensuring that the model was 'scalable':

Promoting a model adapted to the local context

The model adopted is simple and enables wide geographic coverage with a large number of small dairies spread across Haiti, rather than a smaller number of bigger, central units that would be inaccessible due to poor infrastructure. It enables easy collection of milk, while at the same time developing a national brand, providing central support to marketing and economies of scale. The centralization of bulk purchasing (for example, of glass bottles) allows individual dairies to save on costs and to access inputs that would otherwise be unaffordable.

Developing diverse and profitable products

Production and marketing strategies aim to increase the market potential of Lèt Agogo products through diversification and by providing products adapted to local tastes. Winning a government contract to provide milk to schools provided a major boost to profits. Before the earthquake, schools in areas close to the dairies purchased between 1,500 and 5,000 bottles of milk each week, depending on the size of the school. The retail price of the milk allows a healthy profit margin of 10 gourdes ($0.27) per bottle for each dairy. While most dairies produce sterilized milk and yoghurt, others are producing cheese and pasteurized milk. An analysis of the most profitable products will drive future decisions on product focus.[12]

Box 5.2 Building self-reliant producers' associations

Central to the Lèt Agogo project is the creation of local producer associations based around each dairy. These associations provide sources of support for farmers and knowledge-sharing networks, as a male beneficiary in Bon Repos explains: 'If I have a problem with one of my animals, I now have an organized network of people to consult for help. Now my cows get sick much less frequently, and this lets me produce more milk for sale.'

In Limonade in north-eastern Haiti, where the first Lèt Agogo dairy was set up in 2002, women's associations have given their members training on gender equality, which has helped them as producers and in the home. Many women now know that men need not be the sole decision-makers, and that they should play an important part in this process as well.

Through the associations, Veterimed has launched a fodder species improvement programme and has encouraged the building of wells to improve the quality and availability of food and water for livestock. It has also provided veterinary services.

Veterimed's long-term goal is to hand over control of the Lèt Agogo dairies to producer associations, and so it also provides associations with training on legal matters, organizational management and administration, and other topics, with the aim of creating self-sustaining entities that can assume full management of dairies within two to four years of their launch. So far, however, only one of the 13 current dairies, the one in Limonade, has moved to producer-led ownership. The plan is to develop the governance structure of Lèt Agogo/Veterimed, so that the others should follow during the life of the project.

Working across the market system

Oxfam GB and Veterimed aim to ensure that Lèt Agogo is able to work across the whole dairy value chain, from small-scale producers to private sector outlets (such as supermarkets), while also supporting access to services and lobbying to improve the enabling environment.

For example, in order to support access to services, Veterimed has mobilized resources from Intervet, a national network of veterinary workers,[13] and has signed a memorandum of understanding with the Ministry of Agriculture which provides guidance to farmers' organizations affiliated to different dairies. Oxfam is also supporting the Central Unit to provide business development services to the dairies. In order to develop a more constructive policy environment, Veterimed's strategy is to work with producers' associations and with organizations such as NGO Agropresse, a national agricultural information agency.[14] This aims to give producers a voice and to ensure that policy decisions support the dairy sector.

Creating a national network

The Lèt Agogo brand is currently managed by Veterimed. The plan is to integrate the dairies into a national-level network of franchises run by producer associations, helping them to access vital inputs (including all the raw materials that the dairies require – bottles, bottle tops, labels, and so on) and to enter local markets using modern marketing techniques. The final make-up of the network will include representation from the dairy boards, farmers' associations, and the Central Unit. Once this structure has been consolidated, brand management will be transferred to the network itself. The objective is to develop an ownership model that includes small-scale farmers in decision-making and profit-sharing structures.

Supporting the business to grow

Lèt Agogo began in 1999, and in 2008 Oxfam conducted an evaluation of the model's impact. This showed that Lèt Agogo producers had seen an increase in milk production of nearly 67 per cent, in comparison with a decline reported by dairy farmers not involved in the project. This was matched by a much larger increase in income for producers supplying Lèt Agogo dairies. The project has now moved into a second phase and is receiving funding from various organizations including Oxfam's new Enterprise Development Programme (EDP), which links businesses in the developing world with providers of finance, skills, and advice.

The second phase aims to expand the dairy network from 13 to 25 units, with support from other donors. In doubling the number of processing sites, Veterimed and Oxfam aim to increase Lèt Agogo's domestic market share from 0.4 per cent to 5 per cent, while benefiting 2,000 rural producer families across Haiti. Oxfam's contribution to this expansion will be complementary to any budget requested for the recovery of Lèt Agogo's capacity following the 2010 earthquake.

The impact of the earthquake on the programme was great. The Veterimed office in Port-au-Prince was destroyed and two members of staff were killed. Two local dairies were also severely damaged and 45,000 bottles were broken. Farmers' time and financial resources were strained by their hosting of displaced people, and some affected buildings were looted.

Veterimed is working with other partners to rebuild its assets and implement its business plan. At present, the network has 11 functioning dairies, two severely damaged, and five in the process of joining the network. The second phase will eventually be delivered by building on the existing model by:

- **Supporting producers:** This includes technical training on improved animal husbandry, provision of animals and supplies, and other activities. Specifically, Oxfam is contributing funding for the distribution of cows to female producers.
- **Strengthening producer associations:** The project will provide capacity-building training for new and current producer associations. It is anticipated that by the end of the project the associations will be capable of managing all aspects of milk production and processing themselves.
- **Developing dairies:** New dairy facilities will be created and the production capacities of existing dairies will be strengthened. Oxfam is providing funding for new dairy equipment, and anticipates that the 13 existing dairies will ultimately have an increased production capacity, while 12 new dairies will each process at least 200 litres of milk each day. Overall, the project team hopes that the network will process approximately 2 m litres of milk per year, with higher milk quality and increased storage capacity to meet new client demand.
- **Enhancement of dairy marketing:** The marketing and outreach capacity of the Central Unit will be strengthened to help it evolve from being entirely supported by Veterimed and donor funds to becoming a self-funded, independent marketing and supply corporation, with a more diverse client base.
- **Strengthening of dairy sector advocacy:** There are key constraints in the policy environment that inhibit dairy production, therefore lobbying the government to guarantee the pasture land rights of producers, improve their access to state credit, and restrict cheap imports is essential to the success of the model.

The second phase is also committed to improving gender equality in local communities by giving women access to cows, by engaging more women as producers, and by creating women's producer associations in some communities and in others integrating women into mixed associations. The programme aims to ensure that 30 per cent of all new members of producer associations are women and that they will occupy at least 40 per cent of all leadership positions.

Outcomes and impacts to date

After ten years of the project's existence, the results achieved by Lèt Agogo show that smallholder farmers can be competitive dairy producers and that development of the dairy sector is likely to provide significant opportunities for poor women and men to increase their incomes. Key results to date include the following:

- An innovative dairy model has been developed, using simple technology suited to a context of dispersed milk production and basic infrastructure. There has been steady growth in the number of dairies since the first one opened in 2002 in the city of Limonade in the north-east of the country. By 2003 there were five dairies, in 2004 seven, in 2007 ten, and in 2008 twelve. In 2009 thirteen dairies were operational, with over 1,400 producers supplying them.
- The brand is well positioned at the national level. Franchise investors (from the franchise model being developed as part of the project) who want to use the brand must respect the conditions established by Veterimed. These include a minimum 30 per cent level of producer participation in decision-making structures; the existence of a board of directors to manage each dairy; respect for hygiene and quality standards; use of packaging agreed by the network; and participation in Lèt Agogo marketing campaigns.
- The project has successfully attracted the interest of different stakeholders, such as NGOs, youth associations, women's associations, and individuals to invest in the dairy sector at local level. More and more groups want to invest in the business, either creating new dairies with their own capital or with support from the state and/or donors. Individual investors have also shown interest in participating.
- Lèt Agogo has secured an important institutional customer, supplying the Haitian government's national school canteen programme. The programme is the network's main client, accounting for some 70 per cent of its sales. A key part of the growth plan is to develop the institutional market and to sell, for example, to schools outside the capital. An important side benefit is that drinking Lèt Agogo milk provides schoolchildren with a better source of nutrition than powdered milk.
- Of the turnover of each dairy, 65 per cent goes to reinforce the local economy, 40 per cent to dairy farmers, and 25 per cent to dairy workers. This compares favourably with the average percentage earned by dairy farmers in developing countries.
- There have been concrete results from advocacy initiatives: for example, dairy farmers in the north-east of Haiti have been supported to gain legal title to 1,070 hectares of pastureland. One beneficiary from Terrier Rouge explained: 'The association's work is meaningful to us, because it would be much harder for us to mobilize to find land or water as individuals.'

Efforts have been made to increase women's economic participation. One example is the distribution of cows to women (see Box 5.3). This allows them to become members of producer associations and to earn an income from milk production. Over 120 cows were distributed in one district, Limonade, between 2006 and 2008. This increased female membership of the Limonade producers' association APWOLIM from zero to 32.5 per cent in 2008.

Oxfam and Veterimed have also begun promoting reflection on gender issues within one producer association in Limonade. This aims to build women's confidence and encourage male members of the association to reflect on and question their attitudes and behaviours towards women and to actively support activities that help to reposition the role of women. This activity has been well received by men in Limonade and will be expanded to other organizations in future.

Box 5.3 Mama Boeuf: supporting autonomy for women

Veterimed has been supporting women in Limonade, north-eastern Haiti, through the 'Mama Boeuf' cow distribution scheme. To date, 127 cows have been distributed to women in the area, giving them a valuable asset that increases both their financial position and social status in the eyes of their peers. One recipient, Elise Elbeau, says, 'This was such an amazing gift and an amazing start for me. Owning a cow is everything. I was able to be one of the first to show what women can do. I was a pioneer.'

The scheme has enabled women to contribute financially to their households and to their communities. According to Elise, with a regular supply of milk from her own cow, she can earn 60 gourdes ($1.60) a day, which is enough to send her children to school. She adds that, with the earnings from the cow, she can be secure. Another recipient of a cow, Marie-Thérèse, is now earning an additional 420 gourdes a week ($11). This has enabled her to pay school fees, allowing five of her six children to go to school.

According to Veterimed management consultant Gerard Grandin, the Mama Boeuf scheme has had a far-reaching effect on economic and social relations within the community: 'It has really helped women get more independence. Typically, rearing cows was viewed as a male pursuit, but now women can show the community that they're equally capable of holding this responsibility and earning income.'

Evaluation results also show that over 65 per cent of survey respondents believe that earning an income has boosted women's participation in household decision-making. Focus group participants in Limonade commented on changes in their own households, such as greater respect for women and lower levels of conflict and abuse.

Challenges to overcome

The Lèt Agogo business model has become one of national reference for processing agricultural products. However, much remains to be done to consolidate and replicate the model to enable further scale-up, including:

- **Governance structure:** Lèt Agogo's governance structure needs clari-fication. Each dairy has an administrative council, or board of directors, which should involve some producer representatives, and at network level there is a confederation of producer associations, but this is not well organized. Therefore not all the dairies are well managed and the producers themselves do not have the influence, voice, or representation that was originally envisaged.
- **Participation by producers:** A sizeable problem remains in resolving how to integrate producers into decision-making structures. Grouping producers together helps boost access to resources and training. However, many members feel that they have only limited participation in the management and decision-making structures of their associations. Moreover, some dairies have not yet established producer associations.
- **Capacity of the Central Unit:** The Central Unit, which is currently managed by Veterimed, does not have adequate capacity to deliver high-quality marketing support across the franchise. It is also highly dependent on donor funding and needs to ensure that it can fund itself through the supply of inputs and replication of the franchise model.
- **Brand ownership:** The Lèt Agogo brand has until now been owned and managed by Veterimed. The problem for the NGO is how to hand over the brand to a franchised network of dairies while retaining the ability to enforce compliance with the conditions of brand use. To do this, it will be necessary to strengthen individual dairy managements, the Central Unit, and the structure of the whole network. It will also be necessary to formalize the franchise structure, ensuring that all these actors are represented. The risk of compromising the quality of the brand is extremely high, and therefore Veterimed needs to ensure that quality standards are maintained. One way to do that would be by controlling the use of the brand, where it would be associated only with products that meet a certain standard.
- **Approach to investment:** Currently, funding opportunities to set up new dairies are provided by state support or by NGO grants, with priority given to groups of women or young producers who do not already own cows.[15] This could delay the process of strengthening producer participation in the management of the business at both dairy and network levels. Other funding opportunities might arise from commercial or develop-ment banks, or from other financial institutions. In addition, private investors are seen as key players to be attracted.
- **Diversity of the client base:** Product marketing is critical for the Lèt Agogo network. At present, promotion is done mainly through the Central Unit, and is managed by Veterimed. Products are also marketed at local level by the dairies themselves, although as yet there has been little coherent promotional activity at regional level. Dairies located near large towns sell via supermarkets. The bulk of production (70 per cent of the total) goes to meet the school feeding programme contract.[16]

This situation is risky, however, as the programme is financed mainly by donor funds and must be renewed on an annual basis, with no guarantee of being continued in the future.

- **Buy-in from the private sector:** Supermarkets are an important distribution channel for Lèt Agogo's products, but many supermarkets do not share the business's vision and instead prioritize imports or have purchasing practices that disadvantage Lèt Agogo. Veterimed and the producer associations need to develop appropriate campaigns to increase the interest of key actors such as the supermarket sector to support agricultural development across Haiti.

Conclusion

The state is a key actor for the success of an enterprise like Lèt Agogo. In addition to infrastructure investments – such as road building to connect farmers in rural areas with urban markets – small-scale farmers need access to affordable services and inputs. Ideally, the government should also make use of tariffs and quotas to protect the domestic dairy industry and to support local production. NGOs have a key role to play in mobilizing civil society and in lobbying government to create an environment in which domestic production can flourish.

The development of enterprise activities for small-scale farmers cannot be realized without innovative partnerships between producers, the private sector, and technical support groups. NGOs have an important role to play here in facilitating the process and in mediating between different actors. Their key roles include the provision of advice and technical guidance and support for the development of producer associations. The sharing of learning is also essential.

Despite many challenges, the Lèt Agogo project has already achieved success in several key areas. In scaling up the network and moving to a franchise-based model of collective ownership, the project aims to give greater autonomy to rural communities with ownership of productive, viable, and sustainable enterprises.

After the earthquake

The impacts of the earthquake of 12 January 2010 were huge. It left 238,000 people dead, one million people in emergency shelters, and a further half-million internally displaced. Those affected suffered loss of personal as well as public assets, and their productive activities. Much of the capital Port-au-Prince was destroyed. Damage to infrastructure affected the availability and accessibility of basic services. Destruction of roads restricted movement, supply of food, and other types of trade. Many people lost homes, social capital, and support networks and were left at greater risk of crime and disease, due to the increased prevalence of both.

The Lèt Agogo project also suffered as a result of the earthquake. With support from local NGO Veterimed,[17] the project had created a network of small dairy enterprises, with the objective of increasing the incomes of small farmers. Among the dairies' main clients were government schools, the majority of which were in Port-au-Prince. The earthquake destroyed 80 per cent of the schools in the capital and the rest closed for several months, so the project lost its biggest single customer. Damage to roads in and around Port-au-Prince and other affected areas prevented milk from being distributed more widely and, as there was no way of selling all the milk, some producers had to sell their animals. The majority of the milk produced had been sold to state schools, but the earthquake left the state unable to pay its bills.

Looking to the future, the earthquake has brought into focus some of the project's weaknesses, in particular its dependence on too small a number of clients, the complexity of its procurement and logistics chain, and its vulnerability to late payments from the state. However, strong leadership skills were developed as part of the project. The organizations involved have shown themselves to be capable and efficient, and have been able to react and innovate in the face of such challenging and unexpected events.

In the short term following the earthquake, the partners in the project were able to collaborate with the WFP in developing the framework of an emergency food assistance programme. In the longer term, they are engaged in a process of holding strategic discussions and have agreed to design a new business plan. This will include a comprehensive review of the project's production model as well as its distribution model. The second phase of the project will be developed during 2011 with support from Oxfam.

Notes

1   United Nations Development Programme (2009) 'Human Development Report: Statistics for 2000–07', UNDP.
2   Jean, A. (2008) *The Role of Agriculture in the Economic Development of Haiti: Why Are the Haitian Peasants So Poor?* (Second Edition), Authorhouse.
3   For example, food exports fell from 28 per cent of total exports in 1981 to only 6 per cent in 2004. See http://www.veterimed.org.ht [accessed 17 March 2011].
4   World Food Programme (2008) 'Comprehensive food security and vulnerability analysis (CFVSA) – 2007/2008', available from: http://documents.wfp.org/stellent/groups/public/documents/ena/wfp197128.pdf [accessed 17 March 2011].
5   UNCTAD (2010) 'Rebuilding Haiti: a new approach to international cooperation', UNCTAD Policy Brief, [Online] http://www.unctad.org/templates/Download.asp?docid=13002&lang=1&intItemID=2068 [accessed 17 March 2011].
6   Chancy, M. (2005) 'Identification de creneaux potentiels dans les filières rurales haitiennes', Ministry of Agriculture, Natural Resources, and Rural

Development, available from: http://www.veterimed.org.ht/colloque/ Filieres_elevage_final[1].doc [accessed 17 March 2011].

7 Ibid.

8 Ibid.

9 The biggest sources of dairy imports are the EU and Canada (ibid); *New Agriculturalist* (2008) 'Making a splash: milk for the masses in Haiti', [Online] http://www.new-ag.info/developments/devItem.php?a=589 [accessed 17 March 2011].

10 Oxfam GB (2009) 'Stories of change. Project evaluation peport: Haiti – Lèt Agogo', Oxfam, Oxford, for more information see http://www.veterimed. org.ht [accessed 15 April 2011].

11 Ibid.

12 Cheese is Lèt Agogo's most recent product, and offers several key advantages over liquid products: it requires relatively few inputs; can be stored for long periods without refrigeration; and is a good way to 'store' excess milk supply over longer periods of time, avoiding waste during periods of abundance. Moreover, the average profit margin is higher compared with that of yoghurt: $1.80/lb as compared with $0.02/bottle. Oxfam GB (2009) 'Stories of change', op. cit.

13 Intervet (or Entèvèt) is a national professional organization of rural animal health workers, which Veterimed helped to establish. It has more than 1,000 members across Haiti. For more information (in French), see http:// www.veterimed.org.ht/Entevet.html [accessed 17 March 2011].

14 Agropresse contributes to rural development by disseminating information on lessons learned from projects and opportunities in the agricultural sector. For more information (in French), see http://www.agropressehaiti. org [accessed 17 March 2011].

15 For examples, see Oxfam GB (2009) 'Stories of change', op. cit.

16 For more information on this initiative, see (in French) http://www.veterimed.org.ht/actualite_ecole_lait_2007.htm [accessed 17 March 2011].

17 Established in 1991, Veterimed works to help small-scale Haitian farmers increase their incomes and improve their quality of life through training, research, and technical support. Initially focused on animal health, it now also supports producers' organizations to boost productivity and hence incomes through the use of improved, and sustainable, production methods. For more information, see http://www.veterimed.org.ht [accessed 17 March 2011].

About the author

Luc Saint Vil is Programme Co-ordinator, Haiti, Oxfam GB.

CHAPTER 6

Growing partnerships – private sector working with farmers in Sri Lanka

Gayathri Jayadevan

Plenty Foods, a Sri Lankan agri-based company, has adapted its business model to develop a stronger and more reliable supply base by working with multiple stakeholders, including farmer groups, government, and NGOs, as well as with other businesses. The company's relationship with small farmers benefits both sides: by providing market-based opportunities for producers, Plenty Foods has helped to improve the incomes and livelihoods of poor people in rural areas, while a more secure supply base has helped it, in turn, to achieve an annual growth rate of 30 per cent.

Introduction

Sri Lanka comes near the top of the list of 'Medium Human Development' countries in the 2010 Human Development Report. Its ranking of 91 out of 169 is the highest among South Asian countries.[1] However, between 2000 and 2007, 39.7 per cent of the population were living on less than $2 a day.

In 2006, 85 per cent of the country's population were living in rural areas.[2] Agriculture is a key economic sector in Sri Lanka, accounting for 11.7 per cent of GDP (World Bank, 2007).[3] Smallholder farmers contribute more than 60 per cent of total agricultural production.[4] Most of these smallholders cultivate farms of less than five acres in size (typically between 1.5 and 2.5 acres) and depend largely on subsistence farming for their livelihoods. Smallholder agriculture is the main source of income for more than 25 per cent of Sri Lankans living in poverty.[5]

The Sri Lankan economy has experienced accelerated growth over the past five years, despite shocks such as oil price hikes, the Indian Ocean tsunami disaster of December 2004, and (until May 2009) hostilities between government forces and the rebel LTTE (Liberation Tigers of Tamil Eelam). Growth has averaged 4–5 per cent historically and in 2007 reached 6.5 per cent.[6] The private sector accounts for 85 per cent of GDP,[7] and liberalization has increased private sector participation and competition in the delivery of services, particularly transportation, communications, and financial services.

Private companies have also become increasingly active in the agricultural sector, as buyers, suppliers, and employers. However, 80 per cent of poor smallholders and agricultural workers remain at the lower end of agricultural value chains, and income disparities between the high and low ends of value chains have grown. Nevertheless, corporate approaches to poverty alleviation are changing. Although some remain grounded in traditional philanthropic financial giving or PR-based corporate social responsibility (CSR) activities, others are looking to develop different approaches that address issues of sustainability related to agricultural development. One of these approaches involves harnessing the contribution that small producers and workers can make in developing efficient and sustainable supply chains.

Plenty Foods (Pvt) Ltd is a food processing company that sells cereals and healthy snacks, primarily within the Sri Lankan local market. The company is wholly owned by Ceylon Biscuits Ltd. Its products carry high brand recognition from urban to rural markets, and it has also established a good reputation for community engagement and fair sourcing from small farmers in many of the areas where it operates.

Oxfam GB began working with Plenty Foods in Hambantota in Southern Province in south-east Sri Lanka, where producers were seeking new markets primarily for green gram (mung beans) – a major ingredient in Plenty Foods' products. Plenty Foods meanwhile was seeking new suppliers of raw materials who would reliably deliver quality produce. By working together, Oxfam and Plenty Foods were able to strengthen the capacity of local producers to reliably plan, finance, grow, and deliver the products that the company needed. This approach not only reduces poverty and builds farmer enterprises, but also means that the company has a reliable supply of produce and can increase its efficiency. This chapter explores how private sector engagement can provide a route to scale through the replication of a successful business model developed in partnership with small-scale producers.

Smallholder agriculture in Hambantota

Oxfam has been working in Sri Lanka since 1986, on both long-term development and humanitarian issues. As part of this work, it has supported 40,000 small producers and workers in the paddy, dairy, market garden, tea, and coir sectors. Since 2006, Oxfam has implemented a Small- and Medium-Scale Farming (SMSF) project in Hambantota in Southern Province. More than 80 per cent of Hambantota's population depend directly or indirectly on agriculture for their livelihoods. Production is low and most farmers have limited access to markets and services.

The SMSF project aims to reach 5,000 small-scale farmers across five divisions of Hambantota district. It plans to boost smallholder incomes and women's leadership in agriculture by improving production and enabling access to finance, agricultural extension, and business development services, whilst working with government to ensure that policy relating to land titles is implemented. It adopted a market systems approach of engaging the private sector as a critical entry point.

Identifying a private sector partner – Plenty Foods

In 2005 Oxfam GB carried out a market analysis across Sri Lanka, to move its programmes from post-tsunami recovery towards sustainable livelihoods and to determine which product lines were best suited for value chain development, in four areas of the country. Market gardening to produce various crops, including green gram, emerged as a priority opportunity in southern Sri Lanka. The primary market for these products is domestic, through local private processing and trading companies. There is a growing export market for fresh and processed fruits and vegetables. However, because of poor quality and inadequate and inconsistent supply, farmers and exporters had not been able to respond to this demand.

A strategy was developed that would test a method of leveraging change at scale through attracting private sector investment and through the development of new business models which link smallholders with the private sector and which can be replicated by other companies. In order to identify a suitable partner, the programme mapped companies involved in agricultural value chains, using criteria that included:

- Presence in Hambantota and existing interest in agricultural products;
- Shared values and interest in working with smallholders;
- Potential for impact on smallholders;
- Opportunity for Oxfam to learn from the engagement;
- Type of services offered.

Oxfam drew up a shortlist of private sector companies, and selected Plenty Foods (Pvt) Ltd. Plenty Foods trades, processes, and markets a range of fast-moving agricultural products and in 2007 had a turnover of approximately Rs 1.2 bn ($1.8 m), with 85 per cent of its revenue being derived from domestic sales through a network of 60,000 distribution outlets. The company sells a large proportion of its products back into local communities.

Plenty Foods' business model commits it to working with small farmers and ensuring their sustainability by increasing the profitability of their agricultural activities. It has contractual relationships with its outgrower network of small farmers across Sri Lanka, most of these are direct agreements. This means that it provides these farmers with technical and business services, as well as a guaranteed price at market value or contract price, whichever is higher.

Oxfam was keen to learn from this model, but also wanted to work with Plenty Foods to test a new way of working that would build farmers' independence and capacity, and also strengthen the links between Plenty Foods and government and other service providers.

Sourcing from smallholders was already an important part of Plenty Foods' strategy. However, as a result of the partnership with Oxfam, the company changed its way of working to the 'producer organization' model. This meant moving away from engaging directly with individual farmers to working with farmers' groups and facilitating the provision to them of finance, insurance, and seeds, by service providers such as agricultural researchers, private financial institutions, agricultural researchers, and training providers. By working with smallholder groups or their leaders, Plenty Foods was able to reduce the number of extension workers it had, while also increasing the number of farmers it could reach. An additional benefit was that this stimulated local economic development by building areas of expertise in service provision within communities.

Oxfam and Plenty Foods went on to identify commodities that they would work on together. One of the reasons why green gram was selected was the high level of market demand for it, which had the potential to increase farmers' incomes. A cost/benefit analysis of green gram production was conducted and this showed potential profits of Rs. 43,800 ($960) per acre, a healthy margin for smallholder farmers.[8]

Setting up the agriculture stakeholder steering committees

Oxfam's approach to market analysis highlights the wide range of actors that are involved in the development of a market system and facilitates dialogue and joint planning between farmers' organizations, civil society, and public and private sector actors. In its Sri Lanka programme, a forum to enable this was called the agriculture stakeholder steering committee. This committee was facilitated and set up by Oxfam at the district level for the purpose of planning and co-ordinating agriculture-related activities in the district. The committee comprised the district government agent for the Department of Agriculture and Agrarian Services and the Department of Irrigation, as well as agricultural companies, banks, business development service providers, international and local NGOs, donor agencies funding agricultural development in the district, and leaders of farmers' organizations.

The steering committee plays a crucial role in developing all aspects of the market system and finding solutions to key issues affecting market development. Some of these issues are explored in more detail below, but can be summarized as:

- Availability of inputs for improving production (e.g. finance, high-quality seeds, and training).
- Leveraging investment from other sources so that services can be delivered in a co-ordinated way to meet the needs of smallholder farmers.
- Finding new markets by establishing links with export companies and making information from the Export Development Board (EDB) on quality standards and procedures available at village level.
- Linking with other government services: for example, Oxfam identified opportunities to link with crop insurance and national infrastructure development schemes (e.g. irrigation), and worked with the national Disaster Management Centre to map risks, draw up plans, and develop small-scale mitigation projects.

- Working together to ensure that policy is implemented and working with government to resolve problems around land tenure.
- Engaging stakeholders to facilitate women-friendly services for women farmers. For example, banks have set up joint monitoring and recovery systems in order to reduce the time that women spend in making trips to town to make loan repayments. The Department of Agriculture and Agrarian Services has invested in additional resources such as the training of female agricultural inspectors to support women farmers with advice and timely extension services.

Developing organizations and enterprises

As well as multi-stakeholder forums, farmer organization and the development of small enterprises are key ingredients in Oxfam's market development model. Forming producer organizations not only gives farmers easier access to services and markets but it also increases their negotiating power. The development of small processing enterprises enables producers to diversify markets and income sources, as well as to increase the profits they make from their produce.

The farmers involved in the Hambantota SMSF project, who previously were scattered over wide areas and were engaged in subsistence farming, have now been mobilized into 36 farmer organizations, which manage a number of crop-based collective farm groups. Farmers have their own plots but also work on communal farm plots where they share labour, the costs of production, and profits. Among other functions, the farmer organizations collect produce from members (acting as collection mechanisms to supply to intermediaries including local processors, traders and exporters). They also meet regularly to discuss issues with agricultural instructors, purchasing companies, banks, and the Department of Agriculture. Organizing in this way has also enabled farmers to achieve the necessary quantity and quality required for export markets.

Crop-based steering committees

These have been set up by Plenty Foods within village-level farmer organizations in order to transfer technology, know-how, and farming techniques. The committees meet at the start of the farming season to plan their cultivation and to map out the services they will need for it. A lead farmer is appointed by each committee to liaise with different service providers and with Plenty Foods, in order to access the necessary extension services and inputs. In the Hambantota model, the multi-actor, crop-based steering committees have been used to bring independent service providers into the market system. They have also developed approaches that have built farmers' capacity to access services, and helped them to use their newly gained skills and resources to gain power in a diverse range of markets.

Through work with district-level stakeholder steering committees, Plenty Foods has forward contracts with farmer organizations and a four-way Memorandum of Understanding (MoU) with the Department of Agriculture, financial services providers, and the producer organizations. The MoU clarifies the roles and responsibilities of different service providers to support farmers during cultivation. Financial institutions provide loans, according to need, to support cultivation in three phases: for preparation of the land, for purchasing of inputs, and for harvesting. This reduces interest rates compared with what farmers would have to pay if they borrowed the whole amount right from the beginning.

Through the MoU, the Department of Agriculture has an understanding of when to provide extension services and advisory support. The MoU has helped it to provide agricultural extension services that meet the needs of farmers selling to the private sector for export as well as to local markets.

The farmers' organizations have certain responsibilities, such as acting as key communication and liaison points between the lead farmers and service providers. Plenty Foods commits its resources and its extension field officers for follow-up visits during cultivation and also provides advisory support and business development training; these services are facilitated through the farmers' organizations. The farmers usually pay for these services, which also include record-keeping and organizational management. Farm management techniques and agricultural support services are provided free of charge by Plenty Foods as a means of improving quality and productivity for the products that it buys back from the farmers.

In addition, Plenty Foods has worked extensively with government research institutions to develop new technologies to improve production methods and yields. Smallholders benefit from such practices when links between them and government-supported extension services provided through the Department of Agriculture and Agrarian Services are established and maintained.

In the area of financial services, this new model has had a particular advantage, as the programme has been able to leverage greater investment by working with formal financial institutions such as Seylan Bank and People's Bank, rather than relying on much smaller amounts of capital available from micro-finance institutions or from Plenty Foods itself. Contracts with and recommendations from an established company such as Plenty Foods make formal financial institutions more confident in lending to farmers. The loans, however, are negotiated with farmers themselves and distributed directly to them. Links with the government crop insurance scheme have also reduced risks for farmers in taking out larger loans against agricultural activities. This has enabled them to increase the amounts they borrow, reduce the interest rates payable, expand their production into new markets, and build up a credit history.

A key feature of the model piloted in Hambantota was that Plenty Foods played the role of 'broker', rather than delivering all the services required by farmers as part of a contract farming model. Previously Plenty Foods had trained farmers and delivered financial services itself or through micro-finance NGOs. Farmers did not manage their own access to these services and the costs were deducted from the value of the produce they sold to the company. This was adequate in meeting the needs of Plenty Foods' supply chain, but it was limited in that it built smallholder dependency on the company as sole buyer and service provider. This did not fit with Plenty Foods' ethos of sustainability nor with Oxfam's objective of building farmers' capacity to manage their own risks.

Finding new markets and building capacity

One of Oxfam's objectives in the project was to support farmers to increase their power in markets. In addition to the initial relationship with Plenty Foods, Oxfam facilitated a link between the farmers' groups and other companies, to identify and develop export markets.

The Export Development Board (EDB) played a key role in developing linkages with companies to reach the export market. This was facilitated through Oxfam, which has a presence in the capital Colombo as well as in Hambantota. One of the companies identified for this purpose was CR Exporter, an exporter of pineapples. This company's arrangement with the farmer groups included a buy-back agreement and it also provided them with an advisory service. In addition, farmers were offered export subsidies by the EDB as an incentive. Other linkages were developed with Silvermill, a provider of contract farming and semi-processing opportunities for limes, and Onesh, a company providing drip-irrigation facilities, fertilizers, and seeds. These companies have now entered into formal contracts with small farmer groups.

To assist with forward planning, the companies develop cultivation plans with farmer organizations early in the year, depending on market demand for produce. While information about export procedures and quality requirements is available via the EDB, this is not readily accessible at village level, so the steering committee has an important role to play in facilitating the development of market information systems.

Challenges to the model

Despite the project's success to date, there remain a number of challenges. For Oxfam's project team in Hambantota, sustainability has been a continuing concern. Continuity of business planning is not evident: although the enterprises managed by farmer organizations seem to be convinced of the usefulness of business planning, recognition and support of this work by partner organizations is weak. The project acknowledges this as a major concern and has taken action to provide links with business experts to facilitate business-planning workshops and to provide business advice to farmers' groups.

Box 6.1 Poverty Footprinting

When it works well, economic development benefits both business and society. Yet it is not always easy for companies to see their role in development, or to understand how their operations affect people and communities.

Oxfam developed the Poverty Footprint methodology for companies to assess and understand their effects on society and on people living in poverty. The methodology combines local assessments of livelihood impacts, value chain analysis, and an assessment of economic contributions into one comprehensive approach.

By conducting such a study, companies can also improve their knowledge and understanding of their business risks and opportunities. It will help to inform decisions about a company's allocation of resources, support transparency reporting requirements, and facilitate the development of new strategies for the business.

Oxfam engaged in a short study with Indian company Harian Kissan Bazaar (HKB) to assess the effects of its operations on development. Company staff took part in designing the study and in setting the parameters for the questionnaire. When the researchers asked about the relevance of women to the company's operations, staff did not understand the roles that women played in their supply chains or markets. After some initial questioning of customers, it turned out that women formed more than 70 per cent of the consumer population using HKB's products. The company staff then changed the parameters of the study, did follow-up work to improve their understanding of their operations relative to women, and engaged more directly with female consumers. Critically, these staff members now had a greater immediate understanding of their company and the society in which it operated. At the same time, they bought into the project without the company needing to do any internal PR.[9]

Supplying produce to an export company is a big opportunity for small-scale farmers, but at the same time it poses many challenges. It is essential for small- and medium-size farmers to be aware of exporters' requirements in terms of quantity and quality. Merely promoting buy-back, outsourcing, or sub-contracting systems by raising the awareness of farmers about the opportunity is not sufficient. Farmers must also be advised on market requirements, especially those of exporters, on a regular basis.

For many stakeholders, 'training programmes' are the answer to any problem. However, the project needs to stimulate less conventional ways of thinking and to involve stakeholders in the government and private sectors as facilitators of new solutions, rather than providers. Studies such as Oxfam's 'Poverty Footprint' (see Box 6.1) can support companies to develop practices and business models that benefit smallholders and poor people more broadly.

Some farmer organizations and partners are engaged in the provision of micro-credit, and are not interested in acting as facilitators to support farmers to explore cost-effective financial services from the formal sector. As a result, some farmers are currently accessing loans with annual interest rates as high as 30 per cent through farmer organizations and partners, while formal sector/bank rates are at 8 per cent. This conflict of interest needs to be managed in order to ensure that poor farmers have the right to choose and identify financial services as appropriate and affordable for their livelihood activities.

In addition, as farmers enter new markets, their investment, and therefore their risks, will rise, in particular if they take out bigger loans to expand production. The consequences of these risks are as yet unknown, although the forum for discussing such issues already exists.

Key achievements

In Hambantota, Plenty Foods and the other members of the multi-stakeholder forum established by Oxfam's SMSF project have had a significant impact on smallholder agriculture in the district:

*   Five hundred farmers are members of organizations that sell to Plenty Foods.
*   Women farmers lead 70 per cent of crop management committees. Women leaders have gained confidence and skills in supporting other women farmers and in negotiating with service providers.
*   Most women farmers from women-headed households used to practise rain-fed farming, i.e. they cultivated once a year and relied on work as casual labourers for the rest of the year to earn an income. Because of the secure market linkages provided by Plenty Foods, women farmers have now been able to move from subsistence farming to cultivating all year around, which has provided them with a regular stream of income throughout the year.
*   Women farmers have benefited from Plenty Foods' expertise and innovation in cost-effective farm management practices that have allowed them to cultivate high-yielding, short-term crops that are drought-resistant, thus increasing their productivity and cash flow.
*   Access to high-yielding seeds, the promotion of an agri-business approach, and improved land management practices have led to increases in productivity, from 200–300 kg per acre in 2006 to 300–400 kg per acre in 2007.
*   Information about export and private sector quality standards, combined with technical support, has also enabled producers to meet these standards more effectively.
*   Plenty Foods has also benefited. The company has grown over 30 per cent per year since the programme started in 2006. This can be attributed to the lower costs of procuring green gram from smallholder farmers and

the fact that, by working closely with local farmers, Plenty Foods has been able to reduce the risks of price volatility by predicting volumes, quality, and prices more accurately.

- The introduction of an enterprise culture and business development training has led to producers adopting a much more market-oriented approach to agriculture. The impact of this has increased the incomes of green gram producers by 80 per cent, from an average of Rs 3,000 to Rs 5,400 ($66 to $120).

- By developing their own organization, and accessing a wide range of independently provided services, smallholders have gained increased power in markets, with the ability to negotiate and present their business case to stakeholders. Producers selling to Plenty Foods have a secure market, but they are not solely dependent on a single contract or associated embedded services.

- Thirty-two farmers' groups and enterprises have been formed under the project. Among them, farmers' groups and enterprises led by women have emerged, processing fruit and vegetables and leading to an increase in women's incomes. Although facilitated by Oxfam and not directly related to Plenty Foods' intervention, this was possible only because women producers had increased income due to the guaranteed market and higher prices offered by the company. Engagement with Plenty Foods also meant that farmers could build up a credit history and develop the skills and financing needed to expand and diversify their business.

- Farmers have also gained access to affordable financial services, with loans of Rs 50,000 to Rs 150,000 ($1,100 to $3,300) being taken and repaid over periods of six months to three years.

- The use of 50 acres of government land has been secured for collective farming activities for farmers awaiting licences or for new diversification projects (see Box 6.2).

- The business model has been replicated by Plenty Foods in other districts in Sri Lanka, including Monaragala district in Uva province and in Vavuniya in northern Sri Lanka, where Oxfam facilitated the process.

Box 6.2 Working together to influence government: the land issue

Plenty Foods was not directly involved in land advocacy – rather, the company was part of the agricultural stakeholder steering committee that led on addressing issues relating to land rights in Hambantota district. The company's presence as part of the committee was important in emphasizing the potential for agricultural development in the district and therefore the need for the government to legitimize land ownership for small-scale farmers.

The process for registering and licensing land is complex and slow in Sri Lanka. Most farmers in Hambantota do not have licences or land titles, and so are compelled to work illegally on land that belongs to the Yala National Park. This uncertainty prevents farmers from investing in their land over the long term and impedes production. Also, many service providers could not engage with farmers who were cultivating on illegal land. This was therefore a key issue for the crop-based steering committees to resolve.

Collective pressure from the agriculture stakeholder steering committees had a role in influencing the government to speed up the process of issuing deeds and permits. In collaboration with relevant authorities such as the Land Commission, the Divisional Secretariat, and the Department of Agriculture, mobile service programmes were launched to speed up the provision of licences and deeds. These programmes involved a series of meetings in villages for small-scale farmers to discuss land-related issues with representatives of these authorities. These meetings were initially facilitated by Oxfam.

Following the meetings, mobile clinics were organized, with officials travelling to villages to expedite the issuing of permits and deeds where the applicants had legitimate entitlements. To date, through this programme, 168 farmers have been issued with deeds and 200 farmers with permits. Deeds have been drawn up for farmers who had been resettled by the government on government-allocated land but were living there without deeds. Permits allow farmers to cultivate government land for a small fee.

Achieving scale through private sector engagement: what worked?

Engaging a private sector company enabled Oxfam to leverage benefits for many more farmers than could be reached by directly implementing a service delivery project.

The agriculture stakeholder steering committees have been crucial in ensuring that all market actors work together and in leveraging investment from other actors – critical ingredients in achieving scale. Public–private dialogue on agricultural development has become an established practice in Hambantota as a result of this project. This has spin-off benefits, as it has enabled new joint undertakings between organizations that previously had little or no interaction. The level of trust in smallholders has grown among companies such as banks and they are more willing to give loans, whereas previously communities relied on informal lenders and traders for loans.

Supporting farmer organization and adopting a district- and crop-focused participatory approach has contributed to a strong sense of local ownership. This is demonstrated by the proactive attitude of the steering committee members, who set and implement their own agendas and take responsibility for any issues. This is central to the development of a sustainable model of change.

As well as benefits for farmers, there is a strong business case for the private sector to engage with smallholders in this model. While Plenty Foods has social objectives and a commitment to promoting social development, it is still first and foremost driven by its desire to maximize its own profits. The success of this model is that the business case has been demonstrated for the company to engage with producer groups rather than with individual farmers. Smallholder members of such groups have proved to be cost-effective suppliers, as their produce is cheaper than Indian imports and the partnership model has enabled Plenty Foods to plan production schedules effectively.

Box 6.3 Influencing the wider private sector

One way for Oxfam to scale up the impact of its work is to share practical experiences and learning with others in order to influence their practice. Traditionally NGOs have focused on trying to influence national governments or other NGOs and have often adopted a highly critical, lobbying approach when working with the private sector. While lobbying is still a useful part of Oxfam's approach to changing practices, it is increasingly investing in proactive and practical engagement with private sector organizations to help them align their interest in ethical values with practices that make business sense. Oxfam has worked with companies at all levels to understand how changing business practices can make a positive difference to poor people.

In particular, Oxfam has sought to engage with companies which, like Plenty Foods, purchase goods from smallholders as part of their supply chain. Plenty Foods has been an ally here in speaking to the private sector in its own language. When Oxfam organized the event Business Models for a Better World in London in 2009, Mohan Ratwatte, the Managing Director of Plenty Foods, delivered an engaging presentation on his business model to interested senior managers from a wide range of British and international companies.

Oxfam was also able to bring learning from this to the Food Ingredients Fair in Frankfurt – a leading global forum for the food and beverage industry. A seminar on How to Source from Small Scale Agricultural Producers drew on the Plenty Foods case study in order to demonstrate both the business and social benefits of buying from smallholders.

In addition, the project has enhanced Plenty Foods' brand profile as a socially responsible company that provides nutritious products at reasonable prices. Its close co-operation through producer groups with higher numbers of farmers, who are both suppliers and customers, has strengthened the company's share in the domestic food market.

One of Oxfam's key criteria for scalable interventions is that they become self-replicating, with other actors taking the lead in spreading the approaches developed. In this respect the Hambantota project has already had impact at scale, as the success of the model piloted there has encouraged Plenty Foods to replicate this model in other parts of Sri Lanka where the company is investing in extension services, training centres, and infrastructure.

The company plans to increase its cultivation of locally grown corn, green gram, soya beans, maize, and rice to 12,500 acres and its farmer network to 25,000 farmers across Sri Lanka by 2012. It also intends to access new export markets by working in collaboration with multi-stakeholder forums such as the one established in Hambantota, to replicate benefits that farmers have experienced there. A central part of working with such forums includes sharing learning with other companies and institutions, and this could lead to replication across the country of the benefits that farmers in Hambantota have experienced.

Oxfam has also built its own confidence in working with private companies, to act as a facilitator and intermediary with different private sector actors delivering services directly to producers. One of the lessons learned is that companies can be key allies in influencing the implementation of policy and in making changes in the enabling environment. They may have access to individuals and processes within government that NGOs do not. By developing its relationship with companies, Oxfam was able to achieve a wider impact in Sri Lanka. The model of change can only work where the company sees the real value of developing the smallholder business model, but this can have many benefits for the company, including a more reliable supply chain, greater trust among communities, and enhanced brand value.

Notes

1   UNDP (2010) International Human Development Indicators, [Online] http://hdr.undp.org/en/statistics/ [accessed 17 March 2011].
2   Rural Poverty Portal, Sri Lanka Statistics, [Online] http://www.rural-povertyportal.org/web/guest/country/statistics/tags/sri lanka [accessed 17 March 2011].
3   World Bank (2007) http://devdata.worldbank.org/AAG/lka_aag.pdf [accessed 17 March 2011].
4   Centre for Poverty Analysis, Sri Lanka (2008), see http://www.cepa.lk [accessed 17 March 2011].
5   Ibid.
6   Central Bank (2007), see http://www.cbsl.gov.lk/ [accessed 17 March 2011].
7   Ibid.
8   Cultivation plans drawn up by Plenty Foods and farmer societies in Hambantota.

9   Oxfam (2009) 'Oxfam Poverty Footprint: understanding business contribution to development', [Online] http://www.oxfam.org.uk/resources/policy/private_sector/poverty-footprint-business.html [accessed 17 March 2011].

About the author

Gayathri Jayadevan is Private Sector Advisor, Sri Lanka, Oxfam GB.

CHAPTER 7

Bridging the gap – building a financial services system that serves poor people in Sri Lanka

Gayathri Jayadevan

> *Providing financial services for poor people in Sri Lanka in partnership with financial institutions, donors, service providers, and policy-makers, Oxfam GB is helping to build democratic, inclusive financial systems in Sri Lanka that serve the majority – poor people. Oxfam's indirect, facilitative approach aims to cultivate links between the formal financial services sector and people living in poverty. Poor farmers have taken advantage of affordable credit to start or expand businesses, improving their incomes and helping to lift their families out of poverty. Savings and insurance schemes now give them a cushion against hard times. In particular, opportunities for self-employment have increased the security of women and their status in the community.*

Introduction: the financial services sector

In its programmes in Sri Lanka, Oxfam has targeted marginalized communities affected by conflict and natural disasters, especially the 2004 tsunami. Typically, members of these communities are engaged in low-income activities in agricultural value chains, such as small-scale farming and casual labour. Most small-scale farmers have few or no assets and depend for their livelihoods on subsistence farming, cultivating plots of less than five acres. In particular, Oxfam has worked to facilitate links between the formal financial services sector and people living in poverty, with the aim of strengthening the access of poor communities to a broader world of financial systems, markets, and development.

Over the past 20 years, increasing deregulation of the Sri Lankan economy has led to rapid growth in the financial services sector. Realizing the potential of financial services as an instrument to reduce poverty, the government has streamlined the process for private sector financial institutions to obtain licences to operate, thus increasing competition. Today's financial sector is diverse, with member-based financial institutions such as Cooperative Rural Banks, Samurdhi Bank Societies, and Thrift and Credit Cooperative Societies reaching a large number of clients, with the private sector playing a key role in the provision of services.[1]

However, by international standards the sector operates at a very low level. First, the lack of a national sector policy on micro-finance and the absence of a cohesive regulatory and supervisory structure mean that the performance of micro-finance programmes is poor.

Second, despite the relative maturity of the financial services market, Sri Lanka suffers from a regional bias, with service providers concentrated in Western and Southern provinces while Northern and Eastern provinces are hardly financed at all. In addition, many financial products are of poor quality – not client-driven or cost-effective, and poorly targeted. These problems are serious constraints on the sustainability of the financial services on offer and their accessibility for poor and marginalized communities. Some 60 per cent of the country's population remain unbanked.[2]

Issues such as lack of collateral, weak legal systems, and a risk-averse attitude to enterprise culture create a greater dependency on the expensive informal financial sector.[3] Women in particular face difficulties, as they may not have the time, skills, mobility, or assets to meet the requirements of financial services providers. If married, they may not have control over financial services in their own name.

The new three-tier strategy

Improving access to affordable and sustainable financial services has always been a critical element of Oxfam GB's poverty alleviation strategy in Sri Lanka. From the late 1980s up until 2005, Oxfam implemented micro-credit programmes through NGO partners. In 2005, however, this approach was reviewed and it was realized that the model was very expensive in terms of NGO and donor resources, and ultimately unsustainable. There were a number of key reasons for this:

- Oxfam's relatively small capital base meant that the loans were small and limited in geographical coverage.
- Financial services work was not directly linked to a specific value chain, so borrowers took loans that were not subsequently invested in viable livelihoods activities. Oxfam was not able to provide technical support for all the activities.
- Oxfam's service delivery approach meant that it was viewed as a community benefactor; this encouraged a dependency mentality and a feeling amongst communities that they did not need to repay the loans.
- The approach had not responded to the changing context and in some cases involved significantly higher interest rates compared with subsidized credit channelled through the formal sector.
- Oxfam's position, experience, and focus did not enable it to influence the regulatory framework, even though this was a key reason for the poor performance of micro-credit programmes.

Oxfam also wanted to get away from the idea of segregating the financial needs of poor people and instead work to include them in mainstream financial services. It therefore ceased providing capital for finance, and instead moved to facilitate linkages between poor people, their producer organizations in local value chains, and formal or semi-formal finance providers. Oxfam's new role was to act as a broker, forging partnerships between different stakeholders in the formal and semi-formal financial sector: policy-makers and regulators, private sector financial institutions, civil society organizations, and promoters of financial services.

The programme it developed works on a multi-pronged approach, at national policy, institutional, and community levels, and four linkage models have been piloted in communities. These have engaged all types of financial institution in market-based lending, including commercial financial institutions (international and national banks) and formal micro-finance institutions (MFIs).

In developing this strategy, Oxfam undertook socio-economic analysis to identify the poorest people and their credit needs. Community mobilization helped target the poorest for support and organized producers into groups in order to develop their skills in improved production methods, value addition, and business management. Advisory services and business counselling

Table 7.1 Oxfam GB's three strategic aims in the financial services sector

Level of impact	Aim	Target
Tier I **National policy**	Facilitate an enabling environment for the development of sustainable and responsible financial services in Sri Lanka	• Policy • Regulation • Legislation • Supervision • Donors
Tier II **Financial service providers**	Encourage providers of financial services to adopt market-based approaches and invest in quality support services that integrate poor women and men into the formal financial sector in successful ways	• Commercial banks • Non-bank financial institutions • Co-operatives and credit unions • Finance companies • NGO MFIs
Tier III **Individuals/ producer groups**	Build governance capacities of producer organizations, particularly those managed by women, to access financial services from the formal sector on fair and equitable terms and to help give them a more organized voice on livelihoods issues	• Smallholders/workers • MSMEs • Self-help groups • Farmers' societies • Co-operatives • Women's rural development societies

were provided. Oxfam then carried out a baseline survey to assess overall credit needs by sector and district, and a profitability analysis for each sector to establish the maximum affordable interest rates for enterprises in their first year.

Having identified viable financial services institutions, Oxfam then presented its findings and business plans. It entered into dialogue with a range of different institutions, including government-owned institutions, private for-profit institutions, and private non-profit organizations. Other stakeholders with whom it collaborated included government bodies both at national level (Ministry of Finance, Central Bank of Sri Lanka, Insurance Board of Sri Lanka, Agriculture and Agrarian Insurance Board, Ministry of Agriculture) and district level (District Secretariats, Department of Agriculture and Agrarian Services, Divisional Secretariats). In civil society it entered into dialogue with NGO micro-finance institutions, and in the private sector with the Bankers Association, Federation of Chambers of Commerce and Industry in Sri Lanka, and the Institute of Bankers of Sri Lanka. Bilateral donor engagement included the Asian Development Bank (ADB) and the World Bank, via the Central Bank of Sri Lanka.

How the linkage models work

A range of financial services linkage models have been promoted to support the growth and expansion of the dairy, paddy, market garden, and coir value chains in Sri Lanka. To date, more than 100 producer organizations have been involved in this initiative, mainly women's rural development societies, paddy farmers' societies, dairy co-operatives, and community-based organizations registered under the Department of Agriculture, District Secretariats, or the Department of Cooperative Development.

Linkage model 1:
Bank provides finances, producer organization provides support

During Oxfam's negotiations with banks, profitability analysis highlighted the fact that linking clients directly to banks was a more efficient and cost-effective way for members of producer organizations to receive loans than delivering them via NGOs or MFIs. In this model, clients receive loans directly from banks, and producer organizations provide training, technical support, and monitoring. The producer organizations receive an administration fee from banks for monitoring work, which provides them with income. Such monitoring involves the banks working with the producer organizations to build rapport with clients and to follow up on repayments to reduce the risk of default. A joint monitoring and recovery system reduces costs and the risk of defaults, as producer organizations are able to pressure their members on repayment if necessary. Decision-making on loans is based entirely on business feasibility and financial sustainability, a factor that has contributed to 100 per cent repayment and has created interest among the banks to replicate the model in other districts. Currently, the model is being used in Hambantota (market garden value chain), Trincomalee (market garden value chain), and Batticaloa (paddy value chain).

Linkage model 2:
Loans procured and disbursed by a producer organization

The delivery of credit through wholesale loans was selected as the appropriate entry point, because of producer organizations' institutional capacity and their legal entitlement to deliver cost-effective financial services. In this model, producer organizations receive wholesale loans from a number of different financial service providers and pool them into a common fund; they then provide smaller loans to members, with an administration fee (which provides income for the producer organizations) added to the interest rate. The business development services provided by producer organizations further support members in terms of financial and business management. These services include training on business planning, financial management, marketing and business counselling by technical experts. They also facilitate access to market information through collaboration with the Chamber of Commerce. This model is being used in Vavuniya (dairy value chain).

Linkage model 3:
Banks provide both finance and training

Finance is provided through donor funding, so interest rates are subsidized and there is a stronger focus on community empowerment and poverty reduction. In this model, clients are linked directly to banks, which also provide business support and training. Communities are empowered to deal with banks directly and to negotiate terms. Ruhunu Development Bank and Sanasa Development Bank practise this model in the coir sector in Southern Province, with funding support from Central Bank of Sri Lanka.

Linkage model 4 (currently in negotiation):
Banks gain outreach by linking with MFIs

This model focuses on partnerships between international banks and MFIs to provide greater access to end-users living in (ex-)conflict areas or in the remotest parts of the country. The model links large funds from international banks with well-established grassroots systems and the skills of local MFIs to accommodate small-value micro-credit transactions.

Box 7.1 Timely support for village entrepreneur

Kandasamy Nirmala, a resident of Satham Nagar in north-eastern Sri Lanka, wanted to run her own business. Her husband's income from fishing was barely enough to meet the family's daily expenses; she needed something that would earn an income while allowing her to stay close to her children at home. She saw an opportunity in the fact that women from her village had to make a three-hour journey into town to buy groceries, and in 2002 she set up a convenience store in a front room of her house, selling groceries, vegetables, and fruits.

However, she faced competition when several other new convenience stores opened in the village. It was difficult for her to compete with their low prices because they sourced their products from a large wholesale market, while she sourced hers from a local wholesaler.

She needed a loan to gain a competitive edge over the other stores. But as a member of a revolving loan scheme, she was only able to borrow up to Rs. 5,000 ($108) at a time, at an interest rate of 10 per cent per month (120 per cent annually). In addition, she could not access money when she needed it, as she had to wait for other members of the scheme to repay their loans.

Kandasamy is also a member of the Satham Nagar Sarvodhayam Society, a community organization supported by Sri Lankan NGO Sarvodhaya and Oxfam GB. The society helped her to negotiate a loan from People's Bank of Rs. 50,000 ($1,077) at an annual interest rate of 8 per cent, and also helped her to access business management training and support from business development service providers such as Sarvodhaya Economic Enterprise Development Services (SEEDS). This enabled her to invest in a wide range of supplies, attracting new customers, doubling her monthly stock turnover, and increasing sales from Rs. 600 to Rs. 1,500 ($13–33) per day.

She says, 'Now I earn enough to contribute equally towards the household expenditure. I have enough money at hand to look after my children's education needs [and] I now have the confidence to negotiate with bank officers and managers.'

Outcomes

To date, the activities of Oxfam and its partners have assisted some 40,000 wage earners and small-scale farmers, 70 per cent of whom are women. Access to financial services has helped farmers to obtain better prices for their produce, as they are no longer obliged to sell back to the traders who provide loans. On average, farmers have seen increases in farm gate prices of up to 200 per cent. Additionally, they have made savings on loan repayments, as interest rates on loans from the formal sector are as low as 8 per cent annually, compared with the 200 per cent per annum charged by the informal sector. Savings on repayments vary depending on the size of the loan.

The reduction in the costs of production due to the lower interest rates and the increase in farm gate prices have both contributed to generating increased profits. On average, farmers have seen their profits rise by 50 per cent over each production period (typically of six months). This in turn has enabled them to make self-investments. They have moved up the agricultural value chain and are able to engage in processing, transportation, and collective marketing schemes. Wage earners no longer need to rely on working as casual labourers to earn an income, but can now lease land to cultivate and undertake farming activities on a permanent basis. Over the past three years, access to formal financial services has enabled more productive small enterprises to be started and existing enterprises to grow and become more profitable in higher-value sectors of the value chain.

Improving access for women

Oxfam's gender analysis, undertaken at the start of the programme, has helped to support women entrepreneurs, and 70 per cent of new enterprises are owned and managed by women. The gender assessment highlighted the particular financial needs of women (both married and single) and the challenges they faced in convincing banks to provide financial services tailored to their needs.

Gender inequities in access and use of financial services differed significantly for married women living with their husbands and for women heading households on their own. In the case of married women, the main issue was about who in the household controlled the credit obtained and the income earned from it.

Men generally assign responsibility for attending regular producer organization meetings to their wives, a tendency compounded by NGO policies that frequently target women for micro-credit programmes as part of their gender equity programming. However, while on paper the loan may have been taken out by the woman in the household, decisions about what loans to take and the incomes earned from them largely remain in the hands of men. So the challenge with married women was to increase their say within the household through gender empowerment programmes, as well as to create spaces for them to engage independently in non-traditional livelihoods activities. Most women are confined to taking care of farm-related activities and have little or no involvement in agricultural markets. Oxfam's programmes have been working to create opportunities for women to engage in collection, processing, and transport activities, and also to promote women's engagement in managing construction-related enterprises such as masonry and carpentry.

Women heading households on their own by default have control over credit and income. However, social norms often prevent them from being recognized as the family's main breadwinner, and male relatives are frequently seen as such. As a result, women's entrepreneurial activities and need for credit may not be taken seriously or given priority. This is a particular issue in the north-east of the country, where years of conflict have meant that women head a significant percentage of households. Traditional cultural and religious beliefs continue to limit women's mobility, acquisition of skills, ability to make decisions, access to information and resources, and ownership of resources. In creating financial services for women, these issues need to be overcome – for example, by putting more emphasis on training and non-financial services, such as technical, financial, business, and marketing services.

Box 7.2 Supporting women to build sustainable businesses

Sharmila Nazreen Jerad lives in Puthukudiyiruppu, near Trincomalee in Eastern Province. For 20 years there were no formal financial services in her village because of the conflict in the region and displacement caused by it. People relied on limited sources of self-finance and on informal financial services, which limited their opportunities to improve their livelihoods and left them vulnerable to environmental and economic shocks.

Sharmila is Secretary of the Puthukudiyiruppu Women's Rural Development Society (WRDS) which, alongside Oxfam and local NGO SEDOT (Social Economic Development Organization in Trincomalee), negotiated with local banks to get them to respond better to the specific needs of rural clients. The banks' main concern was the risk of non-repayment. Using a series of examples, Oxfam and the WRDS demonstrated that loan repayment rates are often higher for women than for men, and also that targeting women encourages female empowerment and financial self-sustainability, while reducing poverty.

As a result, People's Bank has disbursed loans worth Rs. 200,000 ($4,300) to four members of the WRDS and is currently screening more members for loans. Sharmila herself borrowed Rs. 50,000 ($1,077), which she used to buy a cow; People's Bank also provided her with advice on business and livestock management. The cow had a calf and she was soon able to buy another cow from the proceeds of selling milk. She keeps some of the milk for her family and sells the rest, increasing her monthly income by Rs. 4,000 ($86).

In addition to the extra income, Sharmila has saved the money she would previously have spent on milk powder and curd, which amounts to approximately Rs. 1,500 ($33) per month. She says, 'The loan not only increased my income and assets, but gave me increased status in the household and community.'

Her dream, she says, is to establish a milk-bottling factory, which would supply milk to schoolchildren and provide employment for other women in the village. 'Now, with access to safe and efficient financial services, my dream is alive again. I feel these institutions can support me to take advantage of profitable investment opportunities and take risks in the future.'

Some of the recommendations implemented by banks to date include:

- Creating capacity development programmes specifically geared towards the needs of women, e.g. by including confidence-building and issues related to family responsibilities in training programmes.
- Developing approaches to marketing and promoting financial services that are more likely to reach women, e.g. through direct promotion in marketplaces and homes, through service fairs, in the context of radio programmes popular with women, and by presenting women as role models.
- Developing modalities for delivering financial services that enable women to benefit from them, e.g. bringing loan officers to communities instead of requiring women to go to banks, and providing part-time financial management training spread out over longer periods of time.

Wider range of market-based services

Oxfam collaborated with the Federation of Chambers of Commerce and Industry of Sri Lanka on the Business for Peace initiative to foster dialogue with key financial institutions, present evidence of good practice on linkage banking, and support the scale-up of financial services in the conflict-affected areas of northern and eastern Sri Lanka. This has led to a wider range of market-based services being offered, including loans, savings, insurance, and remittance services. For example:

- The Lanka Putra Bank has opened operations in Eastern Province in order to provide services to support post-conflict recovery and local economic development. Its services are supporting displaced and resettled communities to boost their incomes, thus improving their ability to buy daily consumables, send their children to school, build assets, and start up small and micro enterprises.
- In 2008 two insurance companies, Sanasa and Ceylinco, launched new lines of health, life, and production-related insurance services targeting small producers in specific sectors of value chains. These have enhanced the ability of small producers to cope with external shocks and have reduced the need for them to borrow from moneylenders at high interest rates.
- Crop insurance facilitated through the Agriculture and Agrarian Insurance Board has helped producers to reduce the risk of distress selling of assets during disasters.
- Sharia-compliant services have been promoted through financial services providers Amana Takaful and BRAC to support people who follow Islamic principles to benefit equally from linkages with the formal sector.
- Savings deposits with formal banks have provided vulnerable and conflict-affected communities with increased security and greater capacity to make self-investments.

Advocating for a cohesive policy framework

Implementing a facilitative approach has involved a greater focus on networking and lobbying for Oxfam GB in Sri Lanka, and therefore has increased the potential for it to have a wider impact. Involvement in the Business for Peace initiative and in multi-stakeholder forums at district level has provided opportunities to lobby for improved industry practices. Oxfam has facilitated the formation of multi-stakeholder forums in each sector (dairy, paddy, market garden, tea, and coir) with key government institutions, private sector, and civil society organizations. These forums plan and co-ordinate sector development and advocate for better practice. They are attended by government and industry stakeholders, including private sector financial institutions, and have enabled Oxfam to facilitate dialogue on market-driven product development, using evidence from the pilot work it has done in linking farmers with banks. This is helping to leverage public and private finance to deliver commercial services at scale.

This approach has also allowed Oxfam to have a direct influence on policy outcomes in the financial sector. It has been significantly involved in ensuring that the new Microfinance Institutions Act currently in preparation provides a more helpful regulatory framework for the development of pro-poor financial services. The Act, which will be implemented by the Non-Banking Supervision Department of the Central Bank of Sri Lanka, will streamline the delivery of financial services to prevent wastage of aid funding, duplication of funds, and corruption. A second draft of the Act is being prepared by a core committee consisting of the Central Bank, the Monetary Board, and the Ministry of Finance. The draft was expected to be ready for consultation in mid-2010 and to be presented to Parliament at the end of the year.

In the post-conflict context, Oxfam's dialogue has continued with donors (ADB, World Bank, and the Japan International Cooperation Agency), with government, and with private sector financial institutions to introduce policies and practices that result in sustainable financial services for conflict-affected communities, as opposed to relief-driven micro-finance activities that often dry up once the funding comes to an end.

Sustainability

Overall demand for loans for poor clients in the Oxfam programme areas, for both productive and social purposes, ranges between Rs. 300 m and Rs. 500 m ($2.7–4.7 m). Connecting with several different financial institutions in each region has not only increased the chances of people accessing sufficient credit but also guards against the risk of a single institution withdrawing from the region or local market segment, as well as the risk of conflicts developing with a single lender.

Box 7.3 Lobbying for poor people's access to banking

Some commercial banks request assets as collateral, but women and displaced communities seldom own assets or land. The programme was able to lobby for waivers to such requirements on the basis of strong feasibility. In all cases banks agreed to offer loans on the basis of business plans independently verified as viable, with a character reference from a public servant.

In communities where literacy levels are low, Oxfam and its partners lobbied financial institutions to conduct information sessions and provide support to help villagers fill in forms. The programme involved specialist lawyers and business experts who were able to support communities to negotiate terms. The simplified procedures enabled a greater number of poor women and men to be included and to gain access to formal banking.

For communities to access these services, it was not enough for Oxfam merely to provide information to them and to the financial institutions. In order to encourage institutions to service the poorest communities, Oxfam and its partners worked with them to reduce the high risk of defaults and the cost of service delivery. This involved:

- Clearly explaining to communities that the source of funding for the loans was not from NGOs but from formal financial institutions, and that repayment must be taken seriously.
- Making compulsory the purchase of savings insurance for all households taking loans for production facilitated by Oxfam. The cost of insurance was built into the total loan taken and was marketed by the banks as a financial services package.
- Starting with small loan amounts and gradually increasing them, especially in communities with no previous micro-credit history, thus helping to reduce the risk of borrowers being unable to repay loans.

Oxfam's role has been to support producer organizations to develop good governance that will enable them to interact with formal financial institutions on an equal footing. This has involved supporting them to develop risk management systems for facilitating linkages with institutions, developing their capacity in organizational management and administration, and developing systems for disseminating market intelligence, so that they receive up-to-date information which enables them to make informed choices when choosing which financial institutions to engage with.

When selecting accountable and viable financial service providers that were likely to enter rural communities and continue to offer their services there, Oxfam considered the following factors:

- Their products and services were market-based and reflected a commercial approach, rather than philanthropy.
- They were prepared to commit resources towards non-monetary support services, including capacity-building, technical support, and monitoring.

The programme has shown that it is possible to increase the outreach of financial services and to expand services to poor women and men in even the remotest rural areas, through the use of innovative public–private–producer partnerships. As explained by Mr Udayakumar, Divisional Officer of Vellavally DS Division, 'Capital is always available, but the problem is in creating linkages between financial institutions and farmers, who largely remain unorganized in rural areas.'

Working with mainstream financial service providers to develop new linkage models for financial services benefits many more poor people than the direct beneficiaries of Oxfam programmes, since the changes made are embedded in the policies, attitudes, and practices of formal financial institutions. Furthermore, the models piloted have enormous potential to be scaled up and replicated across Sri Lanka.

Box 7.4 A wider range of market-driven services

The facilitative approach has enabled clients to access other services beyond credit. These include bank accounts, remittance services, insurance schemes, and pensions. For example, Oxfam and its partners have held discussions with the Agricultural and Agrarian Board to introduce a crop insurance scheme for chilli cultivation, adding to existing schemes for paddy, maize, onions, and betel. They are now considering insurance schemes for other crops such as aubergine, okra, snake gourd, bitter gourd, and banana. Oxfam has also facilitated access to life insurance through Ceylinco and Amana Takaful and has linked small producers to government pension schemes.

Both Kandasamy Nirmala and Sharmila Nazreen Jerad have taken advantage of such schemes. Kandasamy has deposited savings at the People's Bank and has taken out a life insurance policy with Ceylinco Life Insurance and a pension scheme with the Department of Agriculture. Sharmila has savings accounts with People's Bank and the Bank of Ceylon. She has also taken out life insurance with Ceylinco and livestock insurance through the Department of Livestock Development. She says, 'Usually we sell assets to combat any external shocks, but now I feel more secure against potential losses and have greater capacity to make self-investments.'

In each of the programme districts in Northern and Eastern provinces, Oxfam has supported the District Secretariat and the Federation of Chambers of Commerce to establish District Enterprise Forums (DEFs). These multi-stakeholder forums bring together producer organizations, policy- and decision-makers, donors, service providers such as financial institutions, and various market actors to contribute to local economic development. The forums have enabled the participating financial institutions to showcase evidence of good practices and lessons learned about reaching rural clients, and this has encouraged and motivated others to follow suit.

Oxfam has recently completed a market assessment on the demand and supply of financial services in the post-war context, and its findings are being disseminated via the DEFs in order to influence the way that services are designed and delivered to conflict-affected communities. Additionally, Oxfam has completed a study on the impacts of co-operative lending policies and practices on women members, and this is being used to shape the Department of Cooperative Development's approach to women producers.

Conclusion

The programme highlights the impact that can be achieved with limited resources by adopting a facilitative approach in the provision of financial services. It shows the benefits of partnering with formal financial service providers, where market-based, cost-efficient, demand-driven financial products have improved opportunities for small producers to participate in higher sectors of the value chain on an equal footing.

Engaging financial service providers to deliver innovative business support services has significantly reduced costs and risks to both institutions and communities. The programme demonstrates how important it is to advocate all the way from local to national levels to create an enabling legal and regulatory environment that allows for the development of inclusive and sustainable financial services, which bring benefits to much greater numbers of poor people.

The programme has proved to be an effective tool in freeing people from poverty and increasing their participation in economic and political systems. Over the past three years, poor people have used affordable credit to start new enterprises and expand existing ones. They have taken advantage of increased earnings to improve consumption levels, send their children to school, access health care, and build their assets. Secure, formal savings and insurance services have provided a cushion when families have needed more money for seasonal expenses or in tough times. With the higher interest rates on savings offered by formal financial institutions, families have been able to increase their savings by up to 20 per cent. There are also indications that they have been able to increase their spending on jewellery, which is considered a form of savings in times of distress.

Women in particular have been empowered to participate in making the decisions that most affect their lives. By providing opportunities for self-employment, the programme has significantly increased women's security, autonomy, self-confidence, and status within households and in the wider community.

Oxfam's next step in engaging with this sector is to continue its research and advocacy efforts to highlight the key changes needed and to shape debate by raising awareness externally, so that financial service providers – as well as policy-makers and donors – better understand the multiple ways that they can affect poor people, and so that they are motivated and pressured to improve their policies and practices.

The key changes needed include:

- **The role of government:** an enabling policy and regulatory environment are required to ensure transparency and to encourage the development of innovative, demand-driven financial services targeting poor women and men.
- **The need for a wide range of services for poor rural people:** rising living costs, increased production shifting towards local markets, and the need to expand and rebuild businesses will all lead to a higher demand for loans and financial safety nets in the post-conflict environment. Financial institutions need to introduce innovative business models and business support services in order to expand their services to cater for rural-based micro-, small-, and medium-scale enterprises, not just well-established urban businesses.
- **Reaching women:** financial institutions need more flexible approaches to providing services and products for women to meet their complex needs.
- **The role of donors:** donors need to be sensitive to existing providers of formal and informal financial services when supporting post-conflict programmes. In particular, in-kind and cash grants and transfers can distort the market and delay the development process for formal financial institutions.
- **Public–private partnerships:** the sector needs to adapt modern technologies to introduce scalable models for delivering financial services (such as mobile banking) through private sector partnerships. This can significantly reduce operational costs for banks and transaction costs for poor clients, and can help reach thousands of people in remote parts of the country.

Although the programme is still at an early stage and an independent evaluation will be needed before firm conclusions can be drawn on its success in terms of the sustainability of loan repayments beyond Oxfam's exit, early indications are positive, as the experience of some of the beneficiaries shows (see Box 7.5).

Box 7.5 Making linkage banking work for small-scale producers

In 2006, H.M. Muthubanda, a 67-year-old farmer from Trincomalee, was selected to be a beneficiary of Oxfam's market garden programme, which is being implemented with the assistance of SEDOT and Sarvodaya. In the initial stages of discussion with all those involved in market gardening, farmers identified difficulties in accessing bank loans as one of their biggest challenges.

Oxfam lobbied the banks to join the programme, and the People's Bank agreed to provide Rs. 50,000 ($445) as a loan to the farmers. As many farmers in the area were also facing difficulties due to water shortages, Oxfam facilitated loan schemes for farmers to build wells and purchase watering equipment.

The loan Muthubanda received from the bank enabled him to cultivate an additional two acres of land and increase his income. He says, 'My earnings from the vegetable garden have allowed me to take care of my family. I am looking forward to improving it some more. With the earnings, I am going to fix up my house, build a well, and expand into the poultry trade. I have great plans for the future.'

In Batticaloa district, farmers have suffered due to the negative effects of conflict and displacement and also because of a lack of low-cost capital and irrigation facilities. In 2008, Oxfam established a Steering Committee to oversee the development of the paddy sector, following an in-depth analysis of the industry.

The committee plays an important role in influencing industry practices and policy outcomes in favour of poor farmers. One of its first activities was to improve linkages between farmers and banks in order to increase farmers' access to affordable financial services from the formal sector. To date, over 10,000 paddy farmers have accessed loans ranging from Rs. 7,500 to Rs. 250,000 ($67–$1,335) at 8 per cent per annum.

In 2008, for the first time in the two decades he has been cultivating his paddy fields, Theivanayagam Southanrajah took a loan from People's Bank. 'Before, I was paying 10 per cent interest each month to the money lender, now I pay the bank 8 per cent interest per annum', he says.

Another farmer who received a loan in the same year is Mohaneswary, who proudly explained how she cultivated paddy with her own money. 'I borrowed Rs. 30,000 ($267) from People's Bank. The total interest I paid was only Rs. 860 ($7.65) over a six-month period. After harvesting, I kept the paddy for two months and sold at a higher price of Rs. 2,300 ($20.46) per 65 kg of rice. If I had sold through the local traders I would have received only Rs. 1,200 ($10.67) per bag. My earnings for the last season increased by Rs. 35,000 ($311).'

Today, People's Bank, Hatton National Bank, and the Bank of Ceylon are participating in this programme. 'The banks are very happy, as 100 per cent of the loans taken in the first year have been repaid. The farmers have also recognized the benefits of taking loans from formal institutions. In the first year of the programme 400 loans were taken; in the second year there were 2,000 loan applications', explained Thyiyagarajah Saravanapavan, a programme officer with Oxfam GB in Batticaloa.

Key lessons learned and future priorities

- A managed risk approach is necessary at all levels when facilitating or brokering relationships between different actors. Oxfam's new approach required many of its staff and partners to become more aware of their roles and responsibilities in driving change and to learn how to make convincing business cases to private sector actors to expand and improve their service provision.

- It is a challenging task to bring marginalized farmers' groups with limited reach together with commercially driven banks that are able to deliver services at scale. However, evidence gathered from Oxfam's pilot linkage models has enabled it to present benchmarks for future replication and scale-up.

- In terms of facilitation, it is important to ensure that activities are shaped by an assessment of the overall financial services market system, in order to identify potential and key constraints that need to be overcome.

- It is important for organizations such as Oxfam to remember that their job is to act as a catalyst in stimulating the financial services sector to become more competitive and to grow to service the poorer sections of society, and that engaging and collaborating with different key stakeholders ensures sustainability.

- It is equally important to have a clear exit strategy, so that once the commercially driven activity of market players (both public and private sector) has grown, the facilitator organization can phase out its involvement. Oxfam has focused on developing a Federation of Producer Organizations at district level, which will facilitate members' access to formal financial services while being recognized as a legally credible organization in their industry.

Notes

1   Asian Development Bank (2002) 'The commercialization of microfinance in Sri Lanka', [Online], http://www.adb.org/Documents/Reports/Commercialization_Microfinance/SRI/default.asp [accessed 15 April 2011].

2   The term 'unbanked' refers to people who have no access to financial services (including savings, credit, money transfer, insurance, or pensions) through any type of financial sector such as banks, non-bank financial institutions, financial co-operatives and credit unions, financial companies, or NGOs (World Bank definition).

3   In Sri Lanka, over 80 per cent of poor rural communities rely on substitutes to banking available in the informal sector, such as village-based money lenders and traders who charge higher rates of interest than market rates.

About the author

Gayathri Jayadevan is Private Sector Advisor, Sri Lanka, Oxfam GB.

CHAPTER 8

Effective co-operation – a new role for cotton producer co-ops in Mali

Abdoulaye Dia and Aboubacar Traore

Falling prices and increased privatization had left small-scale cotton producers in Mali facing an uncertain future. In particular, the role of the state-owned cotton company, on which farmers depended for essential services such as inputs and training, had been greatly diminished. In response, a new initiative involving a coalition of partners has helped to build the capacity of producers' co-operatives to provide services to their own members. It has also helped them to build sustainable partnerships with lending institutions and has increased the participation of women farmers in the running of cotton co-ops. It is now hoped that the programme can be rolled out to the whole of Mali's cotton sector.

Introduction

Mali's cotton production grew from 500,000 tonnes a year in 1997 to a record 635,000 tonnes in 2003, making it the largest cotton producer in Africa at that time.[1] Several hundred thousand people living on 200,000 family farms – some 40 per cent of the rural population[2] – depend directly for their livelihoods on cotton production, which often accounts for up to 75 per cent of their cash income.[3] More than 73 per cent of Mali's cotton farmers live below the poverty line of CFA francs (FCFA) 153,310 ($321).[4] Over 150,000 square kilometres of land in the south of the country are planted with cotton.[5] Typically, the crop is grown on small plots of 2–3 hectares in size, along with cereals and the rearing of livestock, using basic ox plough technology and mainly family labour.

In Mali, cotton is a rain-fed crop planted in May to June and harvested from September to December. Cotton prices are set before the growing season; after harvest, the seed cotton is collected by trucks and delivered to the ginning factories of parastatal cotton company Compagnie Malienne pour le Développement des Fibres Textiles (CMDT).[6] The main output of the ginning process is cotton fibre, of which CMDT sells more than 95 per cent on to the world market.[7] Cotton fibre exports account for 25 per cent of Mali's total export revenues, while cotton production accounts for 8 per cent of national GDP. However, less than 5 per cent of the raw fibre product is processed locally. Mali's economy – among the poorest in the world – is thus extremely vulnerable to external price shocks and fluctuations in world cotton prices.

Through cotton production, smallholders have been able to access credit, fertilizers, and other inputs. Alongside cotton, they have increased their production of livestock and cereals, particularly maize, and have invested in wells, schools, and health facilities. However, in some parts of Mali's cotton basin, soil fertility is in decline, population pressure is rising, and little quality land is available – factors that have affected yields and prompted migration to more productive areas or to towns. Prices for conventional cotton on world markets have been volatile and in decline, partly due to the impact of trade-distorting subsidies paid by industrialized and emerging countries to their cotton farmers.[8]

For three decades, Mali's cotton sector had an efficient and coherent structure for the provision of services to smallholders (from the 1960s to the 1990s), but it has been hit by restructuring and by the withdrawal of state support, as well as by repeated crises linked to cotton trading conditions. A boycott of cotton production in Mali in 2000 was a reaction to falling world market prices due to subsidized production in richer countries. Mismanagement at CMDT exacerbated the situation, and the state was left to cover a large budgetary deficit in the sector.

In 2005 the government adopted a new pricing mechanism indexed to world market prices, and CMDT began to withdraw from larger rural development projects to focus more on cotton production. However, implementation of the new mechanism led to worsening conditions for cotton producers. To date, the major constraint faced by the sector has been exchange rate losses against the euro and the US dollar. These repeated crises led to a government review of how agricultural services were provided, including access to credit for small-scale cotton producers, as part of a wider restructuring of the sector.

Until the late 1990s, CMDT provided extension services that promoted increased organization, literacy, and production skills amongst small farmers, as well as developing rural infrastructure and providing rural employment. It provided training and extension services for cotton production and marketing as part of an integrated cotton system, with all the costs borne within the cotton sector itself. Access to services such as credit and input supplies (both for cotton and for other types of agriculture) was guaranteed by cotton cultivation as part of a vertically integrated value chain.

CMDT's extension system was based on a network of agents working with villages grouped into Rural Promotion and Extension Zones (ZAER) and answering to a district head; the district head led a multi-disciplinary team that included agricultural statisticians, logistics staff, promoters for women farmers, animal husbandry advisers, and others.[9] In all, CMDT agents had regular contact with nearly 200,000 small producers, an arrangement corresponding roughly to one agent for every 130 producers.[10]

However, structural adjustment negotiations between Mali's government and the World Bank led to the introduction in 2001 of the Cotton Sector Development Policy (LPDSC), which stipulated the downsizing of the CMDT and proposed a plan for its privatization. The policy also called for liberalization of seed cotton and cottonseed oil markets, as well as an increased role for producer organizations (POs) in the management of the sector. In response, as part of a restructuring of its agricultural extension services, the government introduced initiatives combining free services (a classic model of producer training and extension services provided with no recovery of costs from beneficiaries) and a co-financing system (testing a model with partial recovery of costs). Both these models, however, have shown limitations in terms of efficiency, sustainability, and scale, with benefits for producers falling short of what was anticipated.

Privatization, combined with these other factors, has had a serious impact on cotton production in Mali in recent years. From being the largest cotton producer in sub-Saharan Africa in 2006, it fell to fourth-largest in 2008. Cotton's proportion of the country's exports declined from 25 per cent to 14 per cent over the same period, while Mali's share of the world cotton market shrank from 3 per cent to 1 per cent. Producers meanwhile saw their income from cotton fall by up to 65 per cent.[11]

Privatization and the changing role of producers

Mali's government has traditionally been heavily involved in the cotton sector through its major shareholding in CMDT, providing farmers with subsidized inputs and covering a significant overall deficit in the sector. However, since privatization and the partial liberalization of the sector began in 2001, the company's role has decreased. Input subsidies stopped in 2002–03 and the scope for price support was drastically reduced by the new price mechanism introduced in 2005.

As part of the reform process, village-level POs (village associations) were expected to assume a new status as formal co-operatives and to take on new functions. The 2001 Co-operative Law laid the legal basis for this change in their role, turning village co-operatives into economic enterprises and permitting them to engage in business activities, guarantee loans from financial institutions, offer advisory services to producers, and supply members with agricultural equipment and inputs. These POs are now known as Sociétés Coopératives de Producteurs de Coton (SCPC – Co-operative Societies of Cotton Producers). Since 2005 the sector has seen the formation of 7,177 SCPCs, compared with 4,400 village associations in 2004. However, while operating at village level, the SCPCs do not necessarily involve all the producers of a given village.

Some major initiatives have been launched to support producer groups. For example, from 2001 to 2004, the World Bank-led Programme of Support for Agricultural Services and Producer Organizations (PASAOP)[12] spent FCFA 947 m on 'demand-led' training tailored to the specific needs of producers. This provided training for thousands of members/leaders of POs, as well as supporting technical training events and advice services nationally. Similarly, the Support Project for Cotton Zone Production Systems[13] had a budget of more than FCFA 500 m over three years, and provided direct support in the form of vocational education to all 7,177 POs.

However, both these initiatives lapsed due to a lack of financing. Others succumbed for the same reason, or because the results they achieved did not justify their continuation. Failed initiatives included the first Centre of Service Provision, financed in large part by the cotton marketing system under a project run by the French Development Agency (Agence Française de Développement, or AFD), and the Union of Rural Management Centres network (Union des Centres de Gestion Rurale).

Oxfam's support for producer organizations

In response to this situation of lost support for the newly created co-ops, Oxfam America and Oxfam GB developed a five-year Programme 'Empowering Producers to Secure Livelihoods in Cotton Growing Regions in Mali and West Africa' for cotton producers for the period 2007–11. The programme focused on livelihoods support in the southern cotton-producing regions of Sikasso and Koulikoro in its first two years, extending to other regions from 2009. It initially targeted 25,000 members of primary producer co-operatives for direct production support or capacity-building activities. Approximately 8,250 of these are producers of organic and fair trade cotton in Sikasso (districts of Yanfolila, Bougouni, Kolondieba, and Garalo) and in the zone covered by the OHVN (Office de la Haute Vallée du Niger – a parastatal rural development agency) in Koulikoro.

In particular, the programme's goal is to reduce small farmers' dependence on cotton in gaining access to essential services. Oxfam's analysis showed that POs were being forced to take on new functions – such as the supply of inputs, the negotiation of loans, and the provision of technical advice to members – without necessarily being prepared and able to do this. Producer groups were also obliged to conform to the conditions of the national Co-operative Law, often without a proper understanding of them.

Oxfam conducted a stakeholder analysis at national and sub-regional levels that identified a number of potential partners, including Malian cotton producers' organizations and international NGOs. At the national level, Oxfam chose to work with the Association of Professional Farmer Organizations of Mali (AOPP),[14] which receives technical support from SNV of the Netherlands[15] for conventional cotton production, and the Malian Organic Movement (MoBioM),[16] which is supported by the Swiss NGO Helvetas.[17] At the sub-regional level, Oxfam partnered with the Association of African Cotton Producers (AProCA),[18] which receives technical support from ENDA Prospectives Dialogues Politiques, a Southern NGO based in Senegal.[19]

Drawing on the experience and knowledge of these partners, Oxfam designed a seven-stage intervention strategy aimed at strengthening the capacity of POs as service providers and supporting their development as sustainable rural enterprises. In contrast with traditional agricultural extension models and with other programmes designed for the cotton sector, Oxfam's programme set out to provide advice aimed specifically at developing entrepreneurial performance, and was based on specific needs identified by co-operative members by means of self-evaluation. The programme also explicitly recognized the need for cotton farmers to diversify into other activities, for example engaging in Organic Fair Trade and Better Cotton Initiative markets, and the need for finance to support this, as well as ensuring that the remaining co-operatives producing conventional cotton were able to reach non-traditional markets. Also apparent was the need for innovative means to empower women farmers within the existing co-operative framework, where they were largely marginalized. An essential aspect of the programme was that it would be 'owned' by the POs themselves, who would be directly involved in managing it.

Box 8.1 Women demonstrating leadership

Awa Boire was the first woman vice-president of the Badjébougou co-operative. When the co-op renewed its board in 2009, women who were literate – and who often had a stronger knowledge of the Co-operative Law than men – stepped up to take on influential positions. Awa was elected as vice-president, along with nine other women who gained positions including treasurer and members responsible for organization, conflict, foreign relations, and provisioning. There are now ten women on the board of directors, accounting for half of its membership.

The Badjébougou co-op chose 8 March 2010, International Women's Day, to challenge Wacoro's communal authorities to show more commitment to – and investment in – adult literacy and other local needs in the communal Plan for Economic, Social and Cultural Development (PDESC). Based on their needs, women co-op members sent 15 credit applications for the purchase of farm equipment under the development plan to the micro-finance institution Kafo Jiginew.[20] Women sent more project files than men, due to their strong presence in the co-op's decision-making bodies.

The national producers' associations were the logical starting point in addressing the changing role of producer groups. These bodies found themselves filling the institutional void left by the withdrawal of CMDT, as POs attempted to supply services to members without first having their own capacities strengthened. The different organizations had different areas of expertise but, between them, they constituted a strong network. Analysis conducted with them focused on how best to reach a large number of POs of different types; how to strengthen POs' capacities to support cotton producers and improve their livelihoods; and how to ensure the long-term sustainability of the intervention.

The programme 'Empowering Producers to Secure Livelihoods in Cotton Growing Regions in Mali and West Africa' has had tangible results. From 2007/08 up to the project's mid-term evaluation in 2009/10, average income per producer increased from FCFA 122,935 ($275) to FCFA 146,175 ($327). With support from MoBioM, organic cotton production increased on average from 300 kg per hectare in 2007 to 407 kg per hectare in 2009/10. There has been a significant increase in women's participation in co-operatives. In first-generation co-operatives producing conventional cotton, women's involvement increased from 1 per cent in 2006/07 to 35 per cent in 2010. In the same period, women's representation on the governing bodies of co-ops increased from zero to 24 per cent. Pilot projects in organic cotton production have seen high levels of participation by women, up to 40 per cent in some areas. A MoBioM guarantee fund of CFA 15 m has been set up to provide women with agricultural equipment.

Implementing the programme

Under Oxfam's cotton sector programme, the umbrella producers' organizations (AOPP, MoBioM, AProCA) have taken on co-ordination and management roles at the national level. At the local level, they have recruited a network of management and training advisers, with AOPP trainers working at village level and MoBioM trainers at district and village levels. For AOPP, the focus is on building institutional knowledge within co-ops to enable them to work as autonomous private enterprises. For MoBioM, there is a greater focus on direct support to production, as organic production is a key diversification activity. Institutional advice and education on climate change adaptation is also included, but currently only for 20 or so co-operatives, as this is still a pilot phase.

In consultation with the village-level SCPCs, the advisers identify at least two members of each co-op to act as promoters, or 'relay stations' for knowledge. These individuals are chosen on criteria such as literacy, analytical ability, and motivation; subsequently they receive training on organizational and management skills.[21] Building a network on this basis makes it possible to reach a large number of POs at a relatively low cost and also ensures that the system is sustainable over the long term.

The programme was designed with a seven-stage plan for implementation. Initially, 215 SCPCs were identified in the programme's intervention zones for conventional cotton production and 60 in organic cotton zones in 2007. By 2009 the total had increased to 403 conventional co-operatives and 73 organic Fair Trade co-ops. Initially, the co-ops' performances, based on a specially designed self-assessment exercise, generally showed a lack of understanding of the Co-operative Law and of the roles and responsibilities of different bodies in the cotton sector, along with a lack of viable economic projects and deficiencies in management and in relationships with financial and technical partners.

Based on weaknesses identified during these self-evaluations, the SCPCs drew up development plans each covering a period of three years. Priorities highlighted in their plans were then translated into projects ready for submission to a financial institution. Each co-op established a commission of members (generally up to five individuals) to draft projects with the help of the co-op adviser; these were then submitted for approval to the co-op General Assembly. By the end of 2009, this stage of the process had been completed by 90 per cent of the participating co-operatives.

Advisers support co-op promoters (the 'relay stations') to draw up a business plan and to secure financing for the activities identified in it. They also monitor the performance of co-op administrators and board members as they apply what they have learned to the everyday running of the organization. Before developing new plans, advisers help co-op leaders to evaluate their economic activities against the yearly goals. Each SCPC undertakes a self-evaluation at the end of each exercise to measure its own performance and to inform the next planning stage.[22]

There have been some challenges in applying this strategy, in particular in the initial identification of implementing partners with sufficient capacity to supply services at the local level. The solution to this was a complementary approach, under which Oxfam chose to support partners with particular competencies and with specific organizational and/or technical skills that would fill the gaps. Each partner in the programme has undertaken to achieve specific results, though the outcomes are evaluated jointly. Another challenge was to ensure the quality of the technical advisers at local level.

Among the programme's strengths is the fact that producer groups, in their self-evaluation sessions, identify the services that they need according to their own priorities. This puts them at the centre of the decision-making process, from initial analysis all the way through to implementation. Having identified areas of weakness, they are assisted with targeted training based on specific needs by the management and training advisers. They in turn have support from the programme's international NGO partners, who provide technical support to the advisers (first-level training of trainers); the advisers then replicate this training with the co-ops' promoters (second-level training of trainers). Finally, the promoters put the training into practice with other co-operative members. This helps to ensure access to quality services at all levels.

Improving women's access to services

A study on gender equity carried out in Mali's cotton zone in 2008[23] showed that the training and advice given to SCPCs have certainly benefited growers – but almost exclusively men, who in Mali are traditionally recognized as cotton producers. Women have not benefited from the support structures available to anywhere near the same extent. Women are traditionally excluded from access to and management of land, and few women were recognized members of SCPCs at the start of the programme. Important decisions about the supply of services are taken by co-operatives' governing bodies, whose members are almost exclusively men. Women are thus excluded from debate and have no opportunity to assert their rights.

In order to increase women's access to services, a 'gender plan' was formulated, with the following objectives:

- To train a cadre of literacy trainers in the cotton zones;
- To provide literacy classes to at least 1,000 women in areas where MoBioM and AOPP are active (to date 1,495 women have been trained);
- To carry out advocacy/lobbying activities to involve women in the SCPCs' decision-making bodies; and
- To facilitate women's access to credit, land title, and agricultural inputs and equipment, through a process of advocacy and negotiation.

There is a critical need to strengthen women's literacy skills and to increase their economic power so that they can become full members of the SCPCs' decision-making bodies. The management and training advisers, as pillars of the programme structure, have received training to increase their sensitivity to the question of women's engagement. In some localities, women have been supported to establish self-help savings groups, which has enabled them to raise funds for their membership fees. The incorporation of women as full members of co-ops guarantees them services to which they would not otherwise have access.

Oxfam and its partners are already engaged in activities aimed at improving literacy to strengthen the capacities of women's associations and at promoting gender equity when planning local development. It is necessary to reinforce women's technical skills to increase their chances of getting onto co-ops' boards as, currently, most key decisions are taken where women are absent.

Guaranteeing access to finance

The programme has set up a guarantee fund to facilitate access to credit for small producers. Credit for cotton production is traditionally guaranteed only via the cotton production system, so the goal of this fund is to facilitate access to finance not guaranteed by cotton, particularly for women producers. It aims to establish a framework of well-trained, capable staff in financial institutions and to foster a relationship of mutual trust between banks and micro-finance institutions (MFIs) and producers. The fund is intended to act as a guarantee against default, reassuring financial institutions and encouraging them to lend to farmers. The plan is for co-operatives to gradually build up a collective 'solidarity fund' which will replace the guarantee fund at the end of the programme, thus ensuring ongoing access to finance.

By the end of the programme, it is intended that POs and financial institutions will have established permanent partnerships based on a relationship of mutual trust. The challenge lies in strengthening the co-operatives' capacities not only in terms of governance and economic management but also in seeking additional markets for non-traditional products. In terms

Box 8.2 Action plan for increasing women's participation

The Gender Pilot Project was developed to ensure greater impact on the lives of women. It aims to build capacity for gender analysis in the cotton sector among actors involved in implementing Oxfam's cotton programme. Another objective is to support actors in preparing and implementing training plans for strengthening female membership of their institutions (AOPP, AproCA, MoBioM, and Union Nationale des Sociétés Coopératives de Producteurs de Coton (UNSCPC – Mali National Cotton Producers Co-operative Union).

Another key outcome that the programme partners would like to see is the strengthening of existing initiatives that support savings and loan schemes self-managed by women. One example of this is 'Saving For Change', an initiative carried out under Oxfam America's leadership, which specifically incorporated literacy training into the system of technical assistance with loans. It also supported the AOPP in implementing its action programme for the promotion of women's participation in the cotton-producing region of Sikasso.

In 2009, 1,528 women learners, aged between 13 and 61 years old and mothers of 4,742 children, participated in intensive adult literacy sessions (REFLECT). In total 1,429 learners, or 94.17 per cent of attendees, took part in the final level test. The results showed 1,163 women (68 per cent) ranked at level 1, which means that they are now able to read, write, and carry out the four basic arithmetic operations. Another 20 per cent were ranked at level 2, which means that they can read and write, but cannot calculate. The remaining 12 per cent were ranked at level 3, which means that they can read only.

During these sessions, women improved their knowledge on topics such as co-operative management, opportunities for economic activity in rural areas, women's access to land, the foundation and management of savings/credit schemes, and the production of organic manure.

In Fana (Dioila district), 319 women from six co-operatives worked together to achieve the following:

- Women's groups in Fantobougou, Djenbenkola, and Yola opened an account with Kafo Jiginew, the MFI managing the programme's guarantee fund.
- Women in Djen started an adult literacy class. — Women's groups in Fantobougou and Yola bought supplies and equipment for their adult literacy classes.
- A women's group in Yola helped pay for children's schooling in the village.
- Women in Yola and Gouana each created their own co-operatives.

of scaling up, the aim is to seek a leveraging effect by asking Mali's leading MFI, Kafo Jiginew, to grant credits for a total amount greater than that of the guarantee fund, which will allow a larger number of beneficiaries to be reached. Currently, Kafo Jiginew has granted four times the value of the fund, which will guarantee credit for up to 300 co-operatives. This alone is still not sufficient, but the results of the initiative will be used to support advocacy work on scaling up targeted at other development agencies and at the government.

Box 8.3 What the programme has achieved to date

At national level, thanks to lobbying activities on specific topics such as cotton pricing, input supply, and literacy programmes targeting women, farmers have obtained subsidies of 50 per cent on fertilizer costs from the government. This has helped to safeguard producers' revenue at a time when cotton prices have fallen by 20 per cent. The 215 primary co-ops involved in the programme are implementing their development plans, each focused on 10 major activities, with an investment budget varying from £67 ($105) to £3,900 ($6,090). A guarantee fund of FCFA 150 mahas been set up with Mali's leading MFI, Kafo Jiginew, to facilitate their access to credit.

The average number of women members of co-operatives has doubled, and in areas where the literacy programme has been conducted, women's membership has increased more than ten-fold. An organic shea butter processing unit and an organic manure production unit employing 40 women have both been set up.

In the organic sector, with the support of Comic Relief funding, MoBioM has sold 60 tonnes of cotton to international retailer TK Maxx.

Sustainability and scalability of services

The programme's approach can be described as one of 'participatory entrepreneurial development'. The system of assistance and training services it supports is aimed at helping local co-operative societies to become viable and totally self-managed enterprises. For the intervention to be sustainable, however, a number of further steps are required to help strengthen the co-ops:

- AOPP needs to expand its technical knowledge (either through SNV's advisers or by employing consultants) in order to train its field advisers. Training content is based on needs identified in the self-assessments conducted by co-operatives, and field advisers are trained progressively according to the stage of development a co-op has reached. This method of institutionalizing knowledge is critical to sustainability.

- Co-operative members should be encouraged to develop a sense of ownership, control, and responsibility for their organizations, by helping them to make decisions in a transparent, democratic, and participatory way.
- Progressive development of socially useful and economically profitable activities should serve as a basis for self-management and for the acquisition by co-op leaders and members of skills in functional literacy, organization, training, marketing, and credit and enterprise management.
- Direct relationships need to be established between local co-operative societies and reliable sources of credit in order to finance co-ops' activities. Profits from these activities can be used to self-finance other economic activities and social infrastructure or services.
- Scaling up requires building a strong knowledge-sharing network of POs to enable them to access technology and the information needed to launch and manage competitive economic activities and defend producers' rights and interests. The national POs can offer such a platform.

At present, the programme's support structure is made up of experienced advisers/trainers with competencies in various aspects of co-operative management. It is planned that this team will gradually withdraw from the day-to-day running of the programme, but AOPP as a national producers' organization will seek internal resources (financial and human) to ensure that the services supplied to the co-ops are maintained. Specifically, AOPP will help set up a local structure to continue the work, possibly linked to the communal-level co-operative associations which could finance advisory support, or as an independent service provider.

The goal by the end of the programme is to ensure that the co-ops have sufficient knowledge and skills to allow them to evolve further on their own. Key to this is the network of promoters established by the programme; currently, about 1,050 promoters are working with the management and training advisers – at least two for each of the 453 co-operatives reached by AOPP and the 73 reached by MoBioM. These promoters may also help to strengthen the capacities of other co-ops not currently involved in the programme, in a spill-over effect. To see that happen, funds should be sought from ongoing programmes at national level. The main constraint on the choice of women as co-op promoters is their low literacy levels, although some co-operatives have chosen solely female promoters (for example, the Nianabougou village co-op in the district of Koutiala has appointed two women as promoters).

At present, one management and training adviser covers on average 14 co-operatives, which means that, in its current form, the programme is able to reach on average 900 producers per adviser. In terms of outreach, this compares favourably with earlier models.[24] The cost of the advisory component of the programme is about FCFA 15,000 ($32) per month per co-operative. This works out at around FCFA 1.50 per kg of cotton produced by a co-operative with an agricultural output of 100 tonnes per year. By comparison, the cost of training traditional extension staff is estimated at about FCFA

5.00 per kg of production. These costs do not include programme management costs in either case.

The programme currently involves a total of 453 conventional co-ops, but the long-term ambition is to reach all the conventional co-operatives in Mali's cotton sector – 7,177 in total. To this end, the programme is developing strategies to influence the decision-making and executive levels of government concerned with providing agricultural services, with the aim of strengthening national-level organizations such as AOPP and MoBioM, and beyond them the national co-ordinating bodies of the 13 member countries of AProCA. In order to ensure expansion, these organizations need to promote this type of intervention as part of a wider national strategy. AOPP alone encompasses 180 farmer organizations and has technical and economic partnerships with seven different donors.

For AOPP and the other implementing partners, mechanisms for scaling up the programme include:

- Influencing the practices of their wider memberships through the development and wider use of tools and approaches promoted by the programme. For instance, AOPP and AProCA are currently seeking to generalize use of the self-evaluation tool across all their member bodies.
- Advocacy for replication of the approach by other development organizations, both private and public. It is hoped that such organizations will be inspired by the lessons of the programme and will replicate it to reach a larger number of POs.

In summary, producers' umbrella organizations are the ideal co-ordinating bodies through which finance can be made available to small producers at a lower cost. They allow service provision to be centralized, permitting economies of scale in services provided to producers. However, it is essential that these bodies are well organized and institutionally solid, so as to guarantee efficient access to services. In addition, scaling up will require other, additional costs that cannot be precisely determined in advance.

Notes

1   Fairtrade Foundation (2006) 'Dougourakoroni cotton producers co-operative, Mali', [Online] http://www.fairtrade.org.uk/producers/cotton/dougourakoroni_cotton_producers.aspx. [accessed 21 March 2011].
2   Ibid.
3   Keïta, M.S. et al. (2002) 'Organisation de la filière cotonnière du Mali: Acteurs, fonctionnement et modes de coordination, Rapport provisoire' ('Organisation of the Cotton Production and Marketing System in Mali: Actors, functioning and modes of co-ordination, Preliminary Report'), RESOCOT-Mali. In 2006 it was estimated that cotton provided an annual cash income of $280 per family.
4   Mali: Cadre Stratégique pour la Croissance et la Réduction de la Pauvreté, November 2006. At that time, $1 = FCFA 477.40.

5 Fairtrade Foundation, op. cit.
6 CMDT owns all of the 17 ginning factories in Mali's cotton-growing zone, and gins and markets all the seed cotton grown. The Government of Mali has a 95 per cent shareholding in the organization, with the remainder held by French company Dagris (Développement des agricultures du Sud, formerly CFDT).
7 Keïta, M.S. et al. (2002) op. cit.
8 Baden, S., Oxfam proposal document for the programme 'Empowering Producers to Secure Livelihoods in Cotton Growing Regions in Mali and West Africa', based on studies carried out during the Make Trade Fair campaign.
9 'Rapport définitif, Appui pour le renforcement des capacités des producteurs' ('Final Report, Support for the Strengthening of Producers' Capacities'), AGROTEC-SPA Rome, June 2002. CMDT agents worked closely with farmers' groups, giving technical advice, delivering inputs, weighing and grading cotton at village level, and recovering credit.
10 The total number of producers in 1999 was 157,877. CMDT (1999) Annual Report 1999, Compagnie Malienne pour le Développement des Fibres Textiles.
11 Observatory of Sustainable Human Development (ODHD) (2009) 'Contribution of cotton to Mali's economic growth', available from: http://www.coton-acp.org/docs/study/Contriution_coton_croissance_econ.pdf [accessed 15 April 2011].
12 Programme d'Appui aux Services Agricoles et aux Organisations de Producteurs.
13 Programme d'Appui aux Systèmes d'Exploitation en Zones Cotonnières.
14 Association des Organisations Professionnelles Paysannes de Mali. AOPP is a federation with a membership of over 180 farmers' organizations, in other sectors as well as the cotton sector.
15 See http://www.snvworld.org/en/countries/mali/Pages/default.aspx [accessed 21 March 2011].
16 Mouvement Biologique Malien. MoBioM is an association of village co-operatives, currently with 73 member organizations, whose members grow cotton and other crops such as mangoes. See http://mobiom.org/ [accessed 21 March 2011].
17 Helvetas 'Organic cotton production in Mali', [Online] http://www.helvetas.org/wEnglish/organic_cotton/info_mali.asp?navtext=Helvetas Projects [accessed 21 March 2011].
18 Association de Producteurs de Coton Africains. AProCA was formed in December 2004 to defend the interests of African cotton producers internationally, in the wake of the 2002 'cotton case' at the WTO. It currently has 15 member countries and a permanent secretariat in Bamako. See http://www.aproca.net/ [accessed 21 March 2011].
19 See http://www.endadiapol.org/ [accessed 21 March 2011].
20 Kafo Jiginew is a Malian micro-finance institution (MFI). Its mission is to offer financial services adapted to the needs of low-income populations, particularly women, who do not have access to formal financial institutions, and help them to develop their income-generating activities.
21 AOPP (2008) 'Capitalisation du processus de sélection des coopératives

beneficiaries' ('Capitalization of the process of selection of beneficiary cooperatives'), AOPP.

22 Mission report of the monitoring/evaluation of the programme, 23 March – 4 April 2009.

23 Bintou Nimaga (2008) 'Etude pour la mise en oeuvre d'un projet pilote genre', ('Study for the implementation of a gender pilot project'), Oxfam Cotton Programme.

24 Report of the programme's baseline study, CERCAD, September 2008. According to the initial reference study, a co-operative has on average 68 producer members

About the authors

Abdoulaye Dia is Cotton Programme Manager, Oxfam GB.
Aboubacar Traore is Programme Performance Manager, Mali, Oxfam GB.

Index

Oxfam GB is a development, relief, and campaigning organization that works with others to find lasting solutions to poverty and suffering around the world. Oxfam GB is a member of Oxfam International.

Oxfam House
John Smith Drive
Cowley
Oxford
OX4 2JY

Tel: +44 (0)1865 472482
E-mail: publish@oxfam.org.uk
www.oxfam.org.uk

Some of the chapters in this book are also available in translation (French and Spanish) as downloadable papers from the Oxfam GB.